Ham and Shortwave Radio for the Electronics Hobbyist

Ham and Shortwave Radio for the Electronics Hobbyist

Stan Gibilisco, W1GV

New York   Chicago   San Francisco
Athens   London   Madrid
Mexico City   Milan   New Delhi
Singapore   Sydney   Toronto

Ham and Shortwave Radio for the Electronics Hobbyist

2 3 4 5 6 7 8 9 0   DOC/DOC   1 2 0 9 8 7 6 5

ISBN 978-0-07-183291-5
MHID   0-07-183291-2

This book is printed on acid-free paper.

Sponsoring Editor
Roger Stewart

Editorial Supervisor
Stephen M. Smith

Production Supervisor
Pamela A. Pelton

Acquisitions Coordinator
Amy Stonebraker

Project Manager
Nancy Dimitry, D&P Editorial Services

Copy Editor
Nancy Dimitry, D&P Editorial Services

Proofreaders
Joe Cavanagh and Don Dimitry,
D&P Editorial Services

Art Director, Cover
Jeff Weeks

Composition
D&P Editorial Services

In memory of Jack,
mentor and friend

About the Author

Stan Gibilisco, an electronics engineer and mathematician, has authored multiple titles for the McGraw-Hill *Demystified* and *Know-It-All* series, along with numerous other technical books and dozens of magazine articles. His work appears in several languages. Stan has been an active Amateur Radio operator since 1966, and operates from his station W1GV in the Black Hills of South Dakota, USA.

Contents

Introduction xiii

1 Hobby Radio: A Technical Overview **1**

Shortwave Radio 1
Amateur (Ham) Radio 3
Citizens Band (CB) Radio 6
Electromagnetic Fields 8
Radio Wave Propagation 10
Morse Code and Radioteletype 16
Voices on Waves 18
Receiver Fundamentals 26
Predetector Stages 30
Detectors 32
Exotic Communications Methods 37

2 Shortwave and Allwave Listening **41**

How I Got into SWL 41
Shortwave Broadcast Bands 44
Sunspots and Shortwave Radio 54
Longwave Radio 57
Dirty Electricity 65
Which Radio's for You? 72

3 Ham Radio Communications Modes **75**

Morse Code 75
Radioteletype 79
Phase-Shift Keying 83
Multiple-Frequency-Shift Keying 88
Amateur Teleprinting over Radio 91
Packet Radio 93
Single Sideband 96
Frequency Modulation 100
Images and Video 103

4 Ham Radio Licenses and Frequencies **109**
Today's License Classes 109
Discontinued License Classes 112
160 Meters 113
80 Meters 115
60 Meters 118
40 Meters 119
30 Meters 122
20 Meters 123
17 Meters 124
15 Meters 126
12 Meters 128
10 Meters 130
6 Meters 132
2 Meters 134
Beyond 2 Meters 135

5 Fixed Ham Stations **139**
Where Will You Put Your Rig? 139
The "Main Radio" 146
Peripheral Equipment 152
Utility-Operated Power Supplies 158
Small Backup Generators 165
Noise, Noise, Noise! 170

6 Mobile and Portable Ham Stations **177**
Mobile Band Options 177
Mobile Power Options 182
Portable Band Options 184
Portable Power Options 190
Stay Safe! 193

7 Ham Antenna Primer **197**
Radiation Resistance 197
Half-Wave Antennas 199
Quarter-Wave Verticals 201
Loops 203
Ground Systems 205
Gain and Directivity 207

Phased Arrays 210
Parasitic Arrays 212
Antennas for UHF and Microwave Frequencies 214
Transmission Lines 217
Stay Safe! 224

8   **Ham Operating Basics** 227
Listen, Listen, Listen! 227
Don't Be a Lid! 229
Signal Reporting 230
Operating in SSB 232
Operating in FM 238
Operating in CW 240
Operating in Non-CW Text Modes 246
Contesting 247
Working DX 250
Rag Chewing 253
Operating with QRP 254
Emergency Preparedness 256

Appendix A   Schematic Symbols 257

Appendix B   Q Signals for Ham Radio 273

Appendix C   Ten-Code Signals for CB Radio 277

**Appendix D   Ten-Code Signals for Police and
                Emergency Personnel** 281

Suggested Additional Reading 287

Index 289

Introduction

This book will help electronics enthusiasts learn about two hobbies that escape the attention of most of today's technophiles: shortwave listening and Amateur (or "ham") Radio. If you have no experience with shortwave listening and want to explore that hobby without getting a ham radio license, this book offers tips and advice. But ultimately, I'd like to see you join the Amateur Radio fraternity. To that end, the book breaks down into eight chapters and four appendixes.

- Chapter 1 explains what radio waves are and how they travel. You'll learn how any sort of data can be "imprinted" on radio waves so that information, and not merely electromagnetic fields, can propagate over great distances.
- In Chap. 2, you'll learn the basics of shortwave listening (which I extend to allwave listening because modern radio receivers cover frequencies that range far above and below the traditional shortwave radio frequencies).
- Chapter 3 defines and explains the communications methods that ham radio operators use. I discuss the assets of the Internet, but emphasize that ham radio will work even when the Internet, and all other utilities, fail.
- Chapter 4 covers the assigned frequency ranges, called bands, that licensed hams use in the United States. Wave-propagation characteristics are discussed, comparing the various bands and offering advice as to those bands best suited for use at different times of the day, year, and sunspot cycle.
- Chapter 5 goes into some detail concerning the selection and installation of equipment for fixed ham radio stations.
- Chapter 6 offers information on how to set up mobile and portable ham radio stations, and relates my own experience operating these station types.
- Chapter 7 describes popular ham radio antennas and transmission lines for various frequencies and station types.

- Chapter 8 details how (and how not) to operate a ham radio station, once it's up and running, using popular voice and digital modes.
- The appendixes show the schematic symbols that electronics hobbyists, radio hams, and engineers use in circuit diagrams, Q signals for ham radio operators, and ten-code signals used by CB, law-enforcement, and emergency personnel.

The idea of communicating with other people over long distances without any artificial infrastructure to carry the signals has mesmerized me ever since I, five years old in the late 1950s, sat at the kitchen table in front of a tube-driven radio and said "Calling! Calling!" into the speaker. No one answered my calls into the ether then, of course, but they did after I got my first ham radio license in 1966. With a little study, a good deal of practice, and a few years of time, the dream morphed into reality! And people answer my signals now when I send CQ, the equivalent of "Calling! Calling!", with a Morse code key or the digital interface of my microprocessor-controlled ham "rig" in my basement "ham shack."

The fact that a few watts' worth of energy waves can squirt from my deck-mounted or vehicle-mounted antenna into the atmosphere, that a minuscule bit of that energy can land in someone else's antenna on the other side of the world, and that we can communicate this way and can keep in contact by means of that energy even if every Internet server on the planet went down—well, the whole business fascinates me to this day. If you know what I mean, if radio holds some magic for you too, then you'll get the most out of this book. But even if you only want to listen to broadcasts on a part of the radio spectrum largely forgotten in the fury of latter-day digital madness, this book has something to offer you.

If you get serious about ham radio, the American Radio Relay League (ARRL), headquartered in Connecticut, offers a bunch of fine publications that go into far more detail on individual aspects of the art than space allows here. These publications can help you get licensed so you can transmit on the ham radio frequencies. Once you're licensed, ARRL publications offer extensive information and guidance involving every imaginable ham radio speciality. The ARRL maintains a website at **www.arrl.org**.

I invite you to look at, and hopefully subscribe to, my YouTube channel. I have uploaded lots of videos involving ham radio and hobby electronics to that site, along with videos supplementing some of my textbooks, a few op-ed blurbs, and even a little mountain fuddy-duddy home-cooking advice (or mischief)!

You might also like my website at **www.sciencewriter.net**. You can e-mail me through the link at that site and let me know what you'd like to see in future editions of this book! If you belong to the ham radio fraternity, you might find me on 14, 18, 21, 24, or 28 MHz doing CW (Morse code) or PSK (phase-shift keying).

73 (Best Regards),
Stan Gibilisco, W1GV

Ham and Shortwave Radio for the Electronics Hobbyist

CHAPTER 1

Hobby Radio:
A Technical Overview

Hobby radio encompasses three activities that anyone with a little spare cash and motivation can enjoy: *shortwave listening* (or *SWL*), *Amateur Radio* (also called *ham radio*), and *Citizens Band radio* (usually called *CB radio*). To some extent, they overlap, although ham radio stands above the other two in terms of "fascination potential" (in my opinion). Let's glance at all three of these hobby radio specialties, and then we'll review the basics of how radios work. I assume that you already know some electricity and electronics theory, and you've picked this book up because you want to take your hobby to new levels. If you need to refresh your memory, I recommend the latest edition of my book *Teach Yourself Electricity and Electronics*.

Shortwave Radio

Starting over 100 years ago, the radio pioneers called the range of frequencies from 3 to 30 megahertz (abbreviated MHz) the *shortwave band*. Today, it's technically called the *high-frequency* (HF) *band*. In a latter-day comparative sense, however, both of these expressions constitute misnomers! The waves are long, as they travel through space, compared to the waves in most wireless communications these days. In addition, this band actually represents quite low frequencies in contemporary terms. But the waves are short, and the frequencies are high, compared to the ones used for radio communications and broadcasting when the terms originated.

In the early 1900s, most wireless communication and broadcasting took place at frequencies below 1.5 MHz (wavelengths of more than 200 meters). Engineers and scientists thought that higher frequencies wouldn't provide good performance in practice, so they all but ignored them as an "electromagnetic wasteland." The vast region of the radio spectrum comprising wavelengths of "200 meters and down," corresponding to frequencies of 1.5 MHz and above, remained subject to the whims and ingenuity of Amateur Radio operators and experimenters, who became known as *hams* because of their flamboyance and supposedly unprofessional character. As things worked out, some hams ended up among the most valuable engineers that the art of communications has ever known.

Within a few years, radio hams discovered that the shortwave frequencies could support long-distance communications and broadcasting on a scale that no one had imagined. In fact, the HF band performed better than the heavily used longer-wavelength frequencies did, allowing reliable contacts spanning thousands of kilometers using transmitters with low or moderate output power. Soon, commercial entities and governments took interest in the shortwave band, and amateurs lost their legal rights to most of it. But hams clung to exclusive operating privileges in small slivers of the shortwave spectrum, thanks in large part to an inventor and ham radio operator named Hiram Maxim, and the HF Amateur Radio bands remain popular to this day.

Shortwave radio still plays a role in international broadcasting, particularly in the developing countries. In technologically advanced nations, most government and business entities have moved their operations to the *very high frequencies* (VHF) from 30 to 300 MHz, the *ultra high frequencies* (UHF) from 300 MHz to 3 GHz, and the *microwave frequencies* above 3 GHz. This ongoing shift has given rise to talk of letting ham radio operators use some additional parts of the shortwave band that they originally discovered.

An HF radio communications receiver, especially one that offers continuous coverage of the spectrum in that general range, is sometimes called a *shortwave receiver*. Most general-coverage receivers function at all frequencies from 1.5 MHz through 30 MHz. Some also operate in the standard broadcast band at 535 kHz to 1.605 MHz. A few receivers, called *allwave* receivers, can function below 535 kHz and into the *longwave radio band*, as well as in bands at frequencies above 30 MHz.

Anyone can build or obtain a shortwave or general-coverage receiver and listen to signals from all around the world. This hobby is called *shortwave listening* (SWL). Millions of people all over the world enjoy it. In the United States, the proliferation of computers and Internet communications has largely overshadowed SWL since the 1980s, and many young Americans grow up ignorant of a realm of broadcasting and communications that still prevails in much of the world.

Various commercially manufactured shortwave receivers exist on the market, ranging in price from under $100 to thousands of dollars. A simple wire receiving antenna, which is all you need to receive the signals, costs practically nothing. Some of the better electronics or hobby stores carry these receivers, along with antenna equipment, for a complete installation. You can also shop around in consumer electronics and Amateur Radio magazines.

> **Tip**
> A shortwave listener in the United States need not obtain a license to receive signals, but in general, a license is required if shortwave radio transmission is contemplated. Shortwave listeners often get interested enough in communications to obtain Amateur Radio licenses, so they can engage in end-to-end-wireless conversations with other radio operators on a global scale, a mode that requires no intervening infrastructure other than the earth's upper atmosphere.

Amateur (Ham) Radio

In most countries, people must obtain government-issued licenses to send messages by means of Amateur Radio. Hundreds of thousands of people have Amateur Radio licenses in the United States. You won't have trouble getting a ham radio operator's license if you know electricity and electronics fundamentals. If you want to get started with this hobby, you can contact your local Amateur Radio club or the *American Radio Relay League* (ARRL) in Newington, Connecticut.

Ham radio operators communicate by talking, sending Morse code, or typing on computers. Typing the text on a computer resembles using the Internet. In fact, some ham radio groups or clubs have set up their own radio networks, and quite a few have "patched" into the Internet as well. Some hams, rather than talking or texting or using Morse code (also called CW for *continuous waves,* even though the waves are broken up into *dots* and *dashes*) on the radio, prefer to experiment with electronic circuits, and sometimes they come up with designs that find their way into commercial and military equipment.

Some hams simply chat about anything that comes to mind (except business matters, which remain illegal to discuss using the ham radio frequencies in the United States). Others like to practice their emergency-communications skills, so that they can serve the public during crises, such as hurricanes, wildfires, earthquakes, or floods. Still others like to venture into the wilderness and talk to people thousands of kilometers away while sitting outdoors under the stars. Radio hams communicate from cars, trucks, trains, boats, aircraft, motorcycles, bicycles, and even on foot.

The simplest ham radio station has a *transceiver* (transmitter/receiver), a microphone, and an antenna. A modest ham radio station fits easily on a desk, and has a footprint roughly the size of a personal desktop computer, keyboard, and monitor. If you want, you can add accessories until your "rig" takes up an entire room in the house or the better part of a cellar, as does my installation shown in Fig. 1-1. You'll also need an antenna of some sort, preferably located outdoors. Figure 1-2 shows my vertical antenna designed to operate at 14 MHz, one of the

FIGURE 1-1 My ham radio station includes a transceiver, an Internet-ready computer, an interface for digital communications, a power output and antenna testing meter, an audio amplifier with four speakers (one shown here, at extreme right) and two displays, one of which is a 46-inch high-definition television set.

FIGURE 1-2 Here's a photo of the antenna for the ham radio station shown in Fig. 1-1. It's made of telescoping sections of aluminum tubing, mounted on a deck railing, and can go up or come down in a couple of minutes to accommodate the rapidly changing weather conditions characteristic of my location.

most popular ham radio frequency bands, also called *20 meters* (approximately the length of the radio waves at that frequency).

Tip

Ham radio is an electronics-intensive and, increasingly, computer-intensive hobby. Radio hams are more likely to own two or more personal computers than are non-hams. Conversely, computer engineers and "power users" are more likely to get interested in ham radio than people who avoid computers. If you use a computer very much, and especially if you're interested in hardware design, you should find it easy to get an Amateur Radio license.

Figure 1-3 is a block diagram of a typical home ham radio station. The computer can serve to network with other hams who own computers, and it functions as a terminal for the transceivers. If desired, the computer can also control the antennas for the station, and can keep a log of all stations that have been contacted. Most modern transceivers can be remotely operated by computer either over the radio or using the Internet.

Figure 1-4 shows a ham radio transceiver for use in a vehicle (upper unit) on the *2-meter* band corresponding to frequencies of 144 to 148 MHz. Two-way radio use from any sort of vehicle (car, truck, boat, train, aircraft, or even bicycle) is called *mobile operation*, and radios especially designed for that application are known as

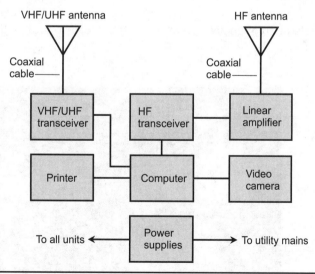

FIGURE 1-3   Block diagram of a well-equipped fixed-location Amateur Radio station. This system has an external *linear amplifier* (a precision radio transmitting amplifier) for generating high-power signals on some frequencies. Abbreviations: HF = high frequency (3 MHz to 30 MHz); VHF = very high frequency (30 MHz to 300 MHz); UHF = ultra high frequency (300 MHz to 3 GHz).

(you guessed it) *mobile radios*. While mobile operation can give you plenty of fun, you should keep in mind the ever-growing possibility that your state or municipality might outlaw it some day if they haven't already. In particular, you might not be allowed to use your transmitter while driving the vehicle.

Warning!   Don't even think about transmitting signals on the Amateur Radio frequencies without getting a license first. If you commit that offense, you can face large fines; in some cases they range in the tens of thousands of dollars. (Yes, the authorities can catch you.) In the United States, several classes of ham radio licenses exist, all of which are issued by the *Federal Communications Commission* (FCC). For complete information, contact the headquarters of the ARRL at 225 Main Street, Newington, CT 06111 or on the Web at

www.arrl.org.

The ARRL publishes excellent books that deal with all subjects relevant to ham radio, as well as up-to-date license exam study materials. The people at ARRL headquarters can tell you the locations of some Amateur Radio clubs near you, where you can meet hams of all "stripes."

Citizens Band (CB) Radio

The *Citizens Radio Service*, also known as *Citizens Band* (CB), is a radio communications and control service. The most familiar CB mode, called *Class D*, operates on 40 channels at frequencies around 27 MHz (corresponding to a wavelength of 11 meters).

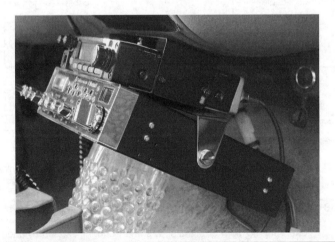

Figure 1-4   Two-way radios mounted under the dashboard of my vehicle. An amateur mobile transceiver rests on top, and a CB mobile transceiver sits underneath it. A plastic cup, wedged underneath the CB radio, keeps both units from wobbling.

Normally, the communications range on the *11-meter band* rarely exceeds 30 kilometers. However, at times of peak sunspot activity, worldwide communication is possible on this band, a feature that brings both good and bad consequences. While you can have fun listening to people who live thousands of kilometers away on a simple radio unit, those same stations can drown out local ones, perhaps the rancher "next door" with whom you want to discuss business (which is legal on CB radio), or the highway patrol officer whom you are trying to reach when your truck breaks down in some remote place where cell phones won't work.

The original 11-meter band had 23 channels, and only *amplitude modulation* (AM) was used. The maximum allowed transmitter power was 5 watts "input to the final amplifier," corresponding to around 3.5 watts of actual radio-frequency (RF) output power. Unlicensed operation was allowed if the power was 100 milliwatts (0.1 watt) or less. Because of the explosion in CB use during the 1970s, the Federal Communications Commission (FCC) increased the number of channels to 40 in the United States, and *single-sideband* (SSB) voice mode was introduced to conserve spectrum space and improve communications reliability. The maximum legal power output was set at 12 watts peak. Licensing requirements were done away with.

The Citizens Radio Service has been "spiced up" by so-called *pirates* and *freebanders*, who use illegal amplifiers and/or operate outside the legal frequency limits of the band. Freeband activity reaches its peak during times of maximum sunspot numbers when long-distance communication can often take place. Some CB operators, legitimate or otherwise, take a conservative route and obtain Amateur Radio licenses to explore the wider horizons offered by that service. They get exposed to a lot more interesting opportunities than the freebanders do, and they can do it all within the scope of the law!

In emergencies, CB radio can save lives. Motorists, hikers, boaters, and small-aircraft pilots use these radios to call for help when stranded. Small CB radio sets, with reduced-size magnetic-mount antennas, are available for motorists, intended for use in emergencies. Some CB operators, in contrast, enjoy the recreational aspect of radio communication. During the 1970s, CB radio became popular among truckers, and remains so to this day, especially in the American South.

Tip

Class-D CB radio equipment doesn't cost much. In fact, some people, having tried CB and deciding that they have no use for it, give their radios away. Check your local newspaper's classified ad section. If you want a new radio, a mobile CB transceiver such as the one shown in Fig. 1-4 (under the smaller ham transceiver) can be had for a hundred dollars or so at Radio Shack stores and various other large general retail outlets. Nearly all of these radios run 4 watts RF power output and use the AM mode.

If you drive through truly remote locations where your cell phone won't work, an HF mobile ham radio set will serve to get you help, even if you have to contact someone hundreds of kilometers away to call the state troopers for you! For these reasons, this book won't deal any further with CB radio. If you're a freebander right now, let me recommend that you get a ham radio license and enjoy radio communications to its fullest. Maybe after reading this book you'll see what I mean!

Electromagnetic Fields

In a radio or television transmitting antenna, electrons move back and forth between the atoms in a metallic wire or tubing. The electrons' velocity constantly changes as they speed up in one direction, slow down, reverse direction, speed up again, and so on. When electrons move, they generate a magnetic field. When electrons accelerate (change speed), they generate a fluctuating magnetic field. When electrons accelerate back and forth at a defined and constant pace, they generate an alternating magnetic field at the same rate, or *frequency*, as that of their motion.

An alternating magnetic field gives rise to an alternating electric field, which in turn spawns another alternating magnetic field, and then another electric field, and so on indefinitely. As this occurs, you get a so-called *electromagnetic* (EM) *field* that can travel through space over vast distances, perpendicular to the electric and magnetic "lines of force," and outward from the energy source (such as your antenna) at the speed of light, approximately 300,000 kilometers per second.

All EM fields have two important properties: the frequency and the wavelength. When you quantify them, you find that they exhibit an inverse relation: as one increases, the other decreases. You can express EM wavelength as the physical distance between any two adjacent points at which either the electric fields or the magnetic fields have identical strength and orientation.

An EM field can have any conceivable frequency, ranging from years per cycle to trillions of cycles per second (hertz). Our sun has a magnetic field that oscillates with a 22-year cycle. Radio waves oscillate at thousands, millions, or billions of hertz. Infrared, visible light, ultraviolet, X rays, and gamma rays comprise EM fields that alternate at many trillions (million millions) of hertz. The wavelength of an EM field can likewise vary over the widest imaginable range, from quadrillions of kilometers to millionths of a millimeter.

Physicists, astronomers, and engineers refer to the entire range of EM wavelengths as the *EM spectrum*. Scientists use logarithmic scales to depict the EM spectrum, as shown in Fig. 1-5A, according to the wavelength in meters (m). The *radio-frequency* (RF) *spectrum*, which includes radio, television, and microwaves, appears expanded in Fig. 1-5B, where the axis is labeled according to frequency. The

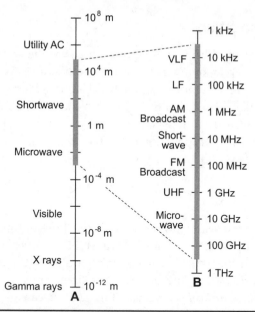

Figure 1-5 Drawing A shows a nomograph of the electromagnetic (EM) spectrum at wavelengths ranging from 10^8 meters (m) down to 10^{-12} m. Each vertical division represents two *orders of magnitude* (two powers of 10). At B, you see a nomograph of the radio-frequency (RF) portion of the spectrum, where each vertical division represents one order of magnitude.

RF spectrum breaks down in *bands* from *very low frequency* (VLF) through *extremely high frequency* (EHF), according to the criteria outlined in Table 1-1. As far as I know, engineers and scientists haven't formally demarcated the exact lower limit of the VLF range; it varies depending on whom you consult. Let's call it 3 kHz.

Table 1-1 Frequency bands in the RF spectrum. Abbreviations: kHz = kilohertz or thousands of hertz; MHz = megahertz or millions of hertz; GHz = gigahertz or billions (thousand-millions) of hertz; m = meters; km = kilometers or thousands of meters; mm = millimeters or thousandths of a meter.

Designation	Frequency Range	Wavelength Range
Very Low Frequency (VLF)	3 kHz to 30 kHz	100 km to 10 km
Low Frequency (LF)	30 kHz to 300 kHz	10 km to 1 km
Medium Frequency (MF)	300 kHz to 3 MHz	1 km to 100 m
High Frequency (HF)	3 MHz to 30 MHz	100 m to 10 m
Very High Frequency (VHF)	30 MHz to 300 MHz	10 m to 1 m
Ultra High Frequency (UHF)	300 MHz to 3 GHz	1 m to 100 mm
Super High Frequency (SHF)	3 GHz to 30 GHz	100 mm to 10 mm
Extremely High Frequency (EHF)	30 GHz to 300 GHz	10 mm to 1 mm

Radio Wave Propagation

The movement of radio waves through space at the speed of light is called *wave propagation*. This phenomenon has fascinated scientists for well over a century, ever since "electromagnetic pioneers," such as Heinrich Hertz, Guglielmo Marconi, and Nikola Tesla, discovered that EM fields can travel over long distances without any humanmade infrastructure, such as wires, cables, optical fibers, or satellites. Let's examine some wave propagation behaviors that affect wireless communications at radio frequencies, and that, if you take up any sort of radio-related hobby, will interest you.

Polarization

You can define the orientation of the electric "lines of force," technically known as *flux lines*, as the *polarization* of an EM wave. If these lines (not real threads or objects of any sort, but theoretical artifacts) run parallel to the earth's surface, you have *horizontal polarization*. If the electric flux lines run perpendicular to the earth's surface, you have *vertical polarization*. Polarization can also have a "slant" that goes at any angle you can imagine.

For Techies Only

In some situations, the electric and magnetic fields rotate as an EM wave travels through space. In that case you have *circular polarization* if the field intensities remain constant all the way around the whole rotational cycle. If the fields are stronger in some planes than in others, that is to say, if their strengths depend from instant to instant on their slant, you have *elliptical polarization*. A circularly or elliptically polarized wave can turn either *clockwise* or *counterclockwise* as you see the wavefronts come toward you. The spin direction is called the *sense* of polarization. Some engineers use the term *right-hand* instead of clockwise and the term *left-hand* instead of counterclockwise.

Line-of-Sight Waves

Radio waves always travel in straight lines unless something makes them bend, bounce, or turn corners. So-called *line-of-sight* propagation can take place when the receiving antenna isn't actually visible from the transmitting antenna because radio waves penetrate nonconducting opaque objects such as trees and frame houses. The line-of-sight wave consists of two components called the *direct wave* and the *reflected wave*.

- In the direct wave, the longest wavelengths are least affected by obstructions. At very low, low, and medium frequencies, direct waves can *diffract* (go around corners in obstructing objects). As the frequency rises, and especially when it gets above 3 MHz or so, obstructions have more blocking effect on direct waves.
- In the reflected wave, the RF energy reflects from the earth's surface and from conducting objects, such as wires and steel girders. The reflected wave always travels farther than the direct wave. The two waves might arrive at the receiving antenna in perfect *phase coincidence* so that they reinforce each other, but usually they don't.

If the direct and reflected waves arrive at the receiving antenna with equal strength but such that one wave lags behind the other by 1/2 cycle, you observe a *dead spot*. The same effect occurs if the two waves arrive inverted in phase with respect to each other (that is, in *phase opposition*). The dead-spot phenomenon is most noticeable at the highest frequencies. At VHF and UHF, an improvement in reception can sometimes result from moving the transmitting or receiving antenna only a few centimeters! In mobile operation, when the transmitter and/or receiver move, multiple dead spots produce rapid, repeated interruptions in the received signal, a phenomenon called *picket fencing*.

Surface Waves

At frequencies below about 10 MHz, the earth's surface conducts alternating current (AC) quite well, so vertically polarized radio waves can follow the surface for hundreds or even thousands of kilometers, with the earth helping to transmit the signals. As you reduce the frequency and increase the wavelength, you observe decreasing *ground loss*, and the waves can travel progressively greater distances by means of *surface-wave propagation*. Horizontally polarized waves don't travel well in this mode, because the conductive surface of the earth in effect shorts out horizontally oriented electric fields.

Tip

At frequencies above about 10 MHz (corresponding to wavelengths shorter than roughly 30 meters), the earth becomes a poor conductor, and surface-wave propagation rarely occurs for distances greater than a few kilometers.

Sky Waves

The upper atmosphere, when influenced by ultraviolet radiation and X rays from the sun, can return radio waves to the earth at certain frequencies. This so-called *ionosphere* has several zones of ionization (regions where the atmosphere's atoms have an electric charge) that occur at fairly constant, predictable altitudes, as shown in Fig. 1-6.

The lowest ionized region is called the *D layer*. It forms at an altitude of about 50 kilometers, and ordinarily exists only on the daylight side of the planet. This layer absorbs radio waves at some frequencies, effectively preventing long-distance radio-wave propagation.

The *E layer*, which forms about 80 kilometers above the surface, exists mainly during the day, although nighttime ionization sometimes occurs. The E layer can provide medium-range radio communication at certain frequencies. Often it forms in "clouds" and can change its character rapidly, hence the origin of the term *sporadic-E propagation*.

At higher altitudes, you find the *F1 layer* and the *F2 layer*. The F1 layer, normally present only on the daylight side of the earth, forms at about 200 kilometers altitude; the F2 layer exists at about 300 kilometers over most, or all, of the earth, on the nighttime side as well as the daylight side. Sometimes, radio enthusiasts ignore the distinction between the F1 and F2 layers, and speak of them together as the *F layer*.

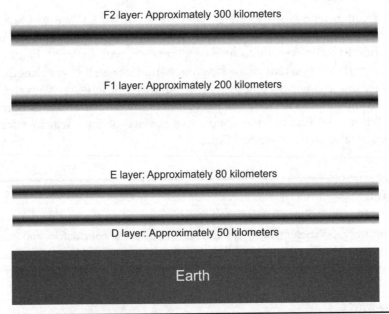

F2 layer: Approximately 300 kilometers

F1 layer: Approximately 200 kilometers

E layer: Approximately 80 kilometers

D layer: Approximately 50 kilometers

Earth

FIGURE 1-6   The ionosphere appears above the earth's surface at various altitudes in layers called D, E, F1, and F2. The most spectacular long-distance shortwave propagation results from the effects of the uppermost layers.

Cool Factoids

"Shortwave" radio communication by means of F-layer propagation can usually be accomplished between any two points on the earth within a certain range of frequencies between 5 MHz and 30 MHz. The bottom of the range, at any given time, is called the *lowest usable frequency* (LUF). The top of the range is called the *maximum usable frequency* (MUF). The exact frequencies of these limits depend on atmospheric conditions, the season of the year, and the locations of the transmitting and receiving station.

Tropospheric Propagation

At frequencies above about 30 MHz (wavelengths shorter than about 10 meters), the lower atmosphere bends radio waves towards the surface. *Tropospheric bending* occurs because the *index of refraction* of air, with respect to radio waves, decreases with altitude. It's the same effect that makes sound waves travel for long distances under certain conditions, such as a calm evening on a big lake. Tropospheric bending can allow you to communicate for hundreds of kilometers, even when the ionosphere will not return waves to the earth.

Another tropospheric-propagation mode is called *tropospheric scatter* or *troposcatter*. This phenomenon takes place because air molecules, dust grains, and water droplets scatter some of the EM field, just as they scatter light rays. You'll observe troposcatter most commonly at VHF and UHF. Troposcatter always occurs to some extent, regardless of the weather conditions. Figure 1-7 shows examples of tropospheric bending and scatter effects.

Ducting, a peculiar type of tropospheric propagation, occurs less often than bending or scatter, but offers more dramatic effects. Ducting takes place when EM waves get "trapped" within a layer of cool, dense air between two layers of warmer air, a phenomenon called the *duct effect*. Like bending, ducting occurs mostly at frequencies above 30 MHz.

Tip
Radio hams often call tropospheric propagation in general, without mention of the specific mode, "tropo."

Auroral Propagation

When the sun gets unusually active, the *Aurora Borealis* ("northern lights") and the *Aurora Australis* ("southern lights") can return radio waves to the earth, facilitating *auroral propagation*. The aurora occur at altitudes of about 60 to 400 kilometers. The effect occurs day and night, even though the "lights" can only be seen visually at

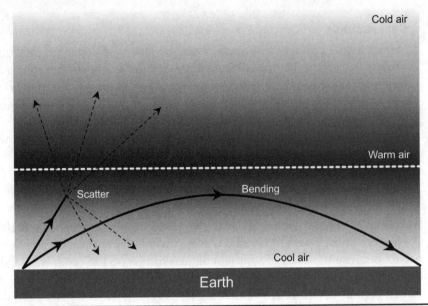

FIGURE 1-7   The earth's lower atmosphere (the troposphere) can affect radio waves at some frequencies. The most common effects are scattering and bending.

night. Theoretically, auroral propagation can occur between any two points on the earth's surface from which the same part of an active aurora region lies on a line of sight. Auroral propagation seldom occurs in the tropics or subtropics, when either the transmitting station or the receiving station is located at a latitude less than 35° north or south of the equator. It's most common in and near the Arctic and Antarctic.

Auroral propagation causes rapid and severe signal *fading* (changes in strength), which renders analog voice and video signals unintelligible. Digital modes work somewhat better, but the signal frequency gets "spread out" or "smeared" over a band several hundred hertz wide as a result of *phase modulation* induced by auroral motion. This "spectral spreading" limits the maximum data transfer rate. For serious users of this mode, Morse code in the form of simple *on/off keying* offers by far the best option. Auroral propagation usually takes place along with poor ionospheric propagation resulting from sudden eruptions called *solar flares* on the sun's surface, so if normal propagation suddenly deteriorates, you might want to check for aurora effects!

Meteor Scatter Propagation

As *meteoroids* from space enter the earth's upper atmosphere to become *meteors*, they produce ionized trails that persist for a fraction of a second up to several seconds. The exact duration of any given *meteor trail* depends on the size of the meteor, its speed, and the angle at which it enters the atmosphere. A single trail rarely lasts long enough to allow transmission of much data. However, during a *meteor shower*, multiple trails can produce almost continuous ionization for a period of hours. Ionized regions of this type can reflect radio waves at certain frequencies. Communications engineers call this effect *meteor scatter propagation*. It can take place at frequencies far above 30 MHz and over distances of up to about 2400 kilometers.

For Techies Only

The maximum communications range that you can expect with meteor scatter propagation depends on the altitude of the ionized trail, and also on the relative positions of the trail, the transmitting station, and the receiving station. On/off keyed Morse code works the best, as does a synchronized digital mode that uses slow, precisely timed, multiple-frequency shifts and goes under the general moniker of *WSJT*. I won't get into the details of that communications scheme here, but if you want to get a good idea of what WSJT is, Google it!

Moonbounce Propagation

Some radio amateurs, blessed with the cash for an elaborate station and a desire for technical adventure, engage in *Earth-moon-earth* (EME) communications, also called *moonbounce*, at VHF and UHF. Successful moonbounce communication requires a sensitive receiver using a specially designed *preamplifier*, a large directional antenna, and a high-power transmitter. The preferred modes are CW and WSJT.

Signal *path loss* presents the main difficulty for anyone who contemplates EME communications. Received EME signals are always weak. High-gain directional antennas must remain constantly aimed at the moon, a requirement that dictates the use of steerable antenna arrays, preferably controlled by computers. The EME *path loss* increases with increasing frequency, but this effect is offset by the more manageable size of high-gain antennas as the wavelength decreases.

Solar noise can pose a problem with moonbounce, as well. Communication becomes most difficult near the time of the new moon, when the moon lies near a line between the earth and the sun because the sun generates strong radio waves (as well as infrared, visible light, ultraviolet, and X rays). Problems can also occur with *cosmic noise* when the moon passes near "bright" regions in the so-called *radio sky*. The constellation *Sagittarius* lies in the direction of the center of the Milky Way galaxy, and EME performance suffers when the moon passes in front of that part of the stellar background because all those stars, like our sun, are gigantic radio noise transmitters!

The moon keeps the same face more or less toward the earth at all times, but some back-and-forth "wobbling" occurs. This motion, called *libration* (not liberation or libation) produces rapid, deep fluctuations in signal strength, a phenomenon known as *libration fading*. The fading becomes more pronounced as the operating frequency increases. It occurs as multiple transmitted radio wavefronts reflect from myriad topographical formations, such as craters and mountains on the moon's surface, whose relative distances constantly change because of the "wobbling." The reflected waves recombine in constantly shifting phases at the receiving antenna, sometimes reinforcing each other to produce a stronger signal, and at other times canceling to produce a deep fade.

Morse Code and Radioteletype

The Morse code is a *binary* (two-state) scheme for sending and receiving messages by electronic means, and it's also the oldest! Engineers call it a binary code because it has only two possible conditions, either "all the way off" or "all the way on." Historically, English-speaking radio and telegraph operators have used two different Morse codes. Nowadays, the commonly used code is the *International Morse Code* or *Continental Code*. In the early days of telegraph, operators used a slightly different code known as the *American Morse code.*

On/Off Keying

In ham radio Morse code, the elements are sometimes called *dots* and *dashes*, where a dot comes over your speaker or headset as a short audio tone and a dash comes through as a longer tone (three times longer, in the ideal case). Speeds are expressed in *words per minute* (wpm).

The Morse code, when transmitted using simple on/off keying, breaks down into an indefinite sequence of digital intervals or *bits* (binary digits), where the signal exists during the whole length of the bit, or else for none of it at all. The key-down (audio tone) condition is called *mark*, and the key-up (silent) condition is called *space*. The dot comprises a mark condition that lasts for one bit; the dash is a mark that lasts for three bits. The spaces between dots and dashes within a single character are one bit long. The spaces between characters are three bits long, and the spaces between words and after sentences are seven bits long.

Table 1-2 shows the International Morse Code characters used in the United States and other countries that have the same alphabet as the English language does. The code characters vary in languages that use other written symbols, such as Cyrillic, Greek, Arabic, Hebrew, Chinese, and Japanese.

Tip
Today's communications devices can function under weak-signal conditions that would frustrate a human operator. But when human operators are involved, the Morse code has proven reliable in getting a message through noise (such as atmospheric "static") because the human ear can easily distinguish between the random nature of noise and the orderly beat of a Morse code signal.

TABLE 1-2 Symbols in the International Morse Code for English and Other Languages That Use the Same Alphabet

Character	Symbol	Character	Symbol
A	. -	W	. - -
B	- . . .	X	- . . -
C	- . - .	Y	- . - -
D	- . .	Z	- - . .
E	.	0	- - - - -
F	. . - .	1	. - - - -
G	- - .	2	. . - - -
H	. . . .	3	. . . - -
I	. .	4	. . . . -
J	. - - -	5	. . . . .
K	- . -	6	- . . . .
L	. - . .	7	- - . . .
M	- -	8	- - - . .
N	- .	9	- - - - .
O	- - -	PERIOD	. - . - . -
P	. - - .	COMMA	- - . . - -
Q	- - . -	HYPHEN	- . . . . -
R	. - .	FORWARD SLASH	- . . - .
S	. . .	COLON	- - - . . .
T	-	SEMICOLON	- . - . - .
U	. . -	PLEASE WAIT	. - . . .
V	. . . -	BREAK OR DOUBLE DASH	- . . . -

Frequency-Shift Keying

You can send digital data faster, and with fewer errors, than Morse code allows if you use *frequency-shift keying* (FSK). In FSK systems, the carrier frequency shifts between mark and space conditions, usually by a few hundred hertz or less. The two most common codes that ham radio operators use with FSK are *Baudot* (pronounced "baw-DOE") and *ASCII* (pronounced "ASK-ee"). The acronym ASCII stands for *American Standard Code for Information Interchange*.

In *radioteletype* (RTTY) FSK systems, a *terminal unit* (TU) converts the digital signals into electrical impulses that operate a teleprinter, or display the characters on a computer screen. The TU also generates the signals necessary to send RTTY when an operator types on a keyboard. A device that sends and receives FSK is sometimes called a *modem*, an acronym that stands for *modulator/demodulator*.

When conditions aren't too bad, FSK works better than on/off Morse-code style keying for data communications because the space signals are specifically identified that way (in other words, their existence is confirmed by an actual signal rather than by silence). A sudden noise burst in an on/off keyed signal can "confuse" a receiver into falsely reading a space as a mark when on/off keying is used, but when the space is positively represented by its own signal, this type of error happens far less often.

Voices on Waves

An audio voice signal comprises frequency elements that lie mostly in the range between 300 Hz and 3 kHz. You can modulate some characteristic of a radio wave, also called a *radio-frequency* (RF) *carrier*, with an audio voice waveform, thereby transmitting the voice information over the airwaves.

Amplitude Modulation

Figure 1-8 shows a simple bipolar-transistor circuit for obtaining *amplitude modulation* (AM). You can imagine this circuit as an RF amplifier for the carrier, with the instantaneous gain (moment-to-moment amplification factor) dependent on the instantaneous audio input amplitude.

The circuit shown in Fig. 1-8 will perform quite well, as long as you don't let the audio input level get too high. If you inject an audio signal that's too strong, you'll end up with *distortion* (nonlinearity) in the transistor, resulting in degraded

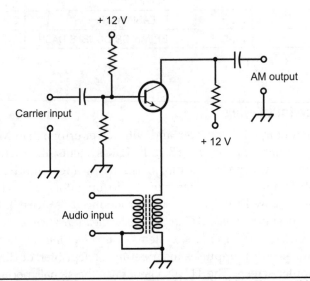

FIGURE 1-8   An amplitude modulator using an NPN bipolar transistor. You can think of this circuit as an amplifier whose gain (amplification factor) depends on the voice input level from instant to instant in time.

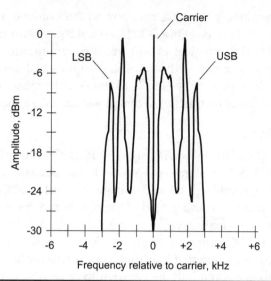

FIGURE 1-9 Spectral display of a typical amplitude-modulated (AM) voice communications signal.

intelligibility (understandability), reduced *circuit efficiency* (ratio of useful power output to total power input), and excessive output signal *bandwidth* (the difference between the highest and lowest signal frequency).

In an AM signal, you can express the modulation extent as a percentage ranging from 0%, representing an *unmodulated carrier*, to 100%, representing the maximum possible modulation you can get without introducing distortion into the signal. In an AM signal modulated at 100%, you'll find that 1/3 of the signal power conveys the voice information, while the carrier wave gobbles up the other 2/3 of the power. That characteristic makes AM a mighty inefficient way to convey information!

Figure 1-9 shows a *spectral display* of a generic AM voice radio signal at a single instant in time. Frequency appears on the horizontal scale, which runs in increments of 1 kHz per division. Amplitude appears on the vertical scale; each division represents ±3 dB of change in signal strength (a doubling or halving of the signal power). The maximum (reference) amplitude of 0 dB represents 1 milliwatt of power, a condition that engineers abbreviate as 0 dBm (0 decibels relative to 1 milliwatt). In a display of this type, the voice data shows up as *sidebands* above and below the carrier frequency. These sidebands are the result of sum and difference signals produced by a mixing effect that takes place in the modulator circuit between the audio and the carrier. The RF energy between −3 kHz and the carrier frequency is called the *lower sideband* (LSB); the RF energy from the carrier frequency to +3 kHz is called the *upper sideband* (USB).

Engineers define the *bandwidth* of any modulated signal as the difference between the maximum and minimum frequencies that the signal contains. In AM, the bandwidth always equals twice the highest audio modulating frequency, assuming that the transmitter is working properly. In the example of Fig. 1-9, all the audio input energy exists at or below 3 kHz, so the bandwidth of the modulated signal equals 6 kHz. That's typical of AM voice communications. In AM broadcasting where music goes along with voices, the audio input energy gets spread over a wider bandwidth, nominally 10 kHz to 20 kHz. The increased bandwidth provides for better *fidelity* (sound quality) so that music doesn't sound muffled in the receiver.

Single Sideband

As mentioned earlier, in an AM signal with 100% modulation, the carrier wave consumes 2/3 of the signal power, and the sidebands exist as mirror-image duplicates that, combined, take advantage of only 1/3 of the signal power.

Suppose that you could get rid of the carrier and one of the sidebands, but still convey all the information you want. In that case, you could get the same "signal bang" with far less transmitter power. Alternatively, you could get a stronger signal for the same amount of transmitter output power. You'd also reduce the signal bandwidth to just about half the bandwidth of an AM signal modulated with the same voice (actually, a little less with the carrier and one sideband completely gone). The resulting *spectrum savings* would allow you to fit more than twice as many signals into a specific range, or *band*, of frequencies, as you could do with AM. During the early twentieth century, communications engineers perfected a way to modify AM signals in precisely this way. They called the new communications mode *single sideband* (SSB), a term which endures to this day.

When you remove the carrier and one of the sidebands from an AM signal, the remaining energy has a spectral display resembling the graph of Fig. 1-10. In this

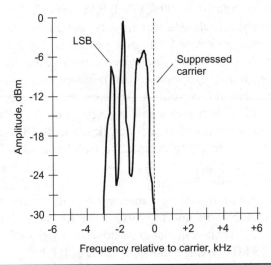

FIGURE 1-10   Spectral display of a hypothetical single-sideband (SSB) voice communications signal, in this case lower sideband (LSB).

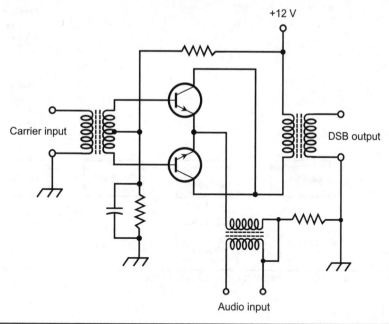

FIGURE 1-11   A balanced modulator using two transistors.

case, the *upper sideband* (USB) energy along with the carrier wave has been eliminated, leaving only the *lower sideband* (LSB) energy. (You could, of course, just as well remove the LSB along with the carrier, leaving only the USB.) Radio hams make extensive use of both LSB and USB on their frequency bands, especially the ones below 30 MHz. They favor LSB at frequencies below 10 MHz, and USB at 10 MHz and above.

Balanced Modulator

You can suppress the carrier in an AM signal using a *balanced modulator*, an amplitude modulator/amplifier using two transistors (or, in the olden days, vacuum tubes) with the inputs and outputs configured as shown in Fig. 1-11. This arrangement cancels out the carrier wave in the output signal, leaving only LSB and USB energy. The balanced modulator produces a *double-sideband suppressed-carrier* (DSBSC) signal, often called simply *double sideband* (DSB). You can suppress one of the sidebands in a subsequent circuit with the addition of a *bandpass filter* to obtain an SSB signal.

Basic SSB Transmitter

Figure 1-12 is a block diagram of a simple SSB transmitter. The RF amplifiers that follow any type of amplitude modulator, including a balanced modulator, must all

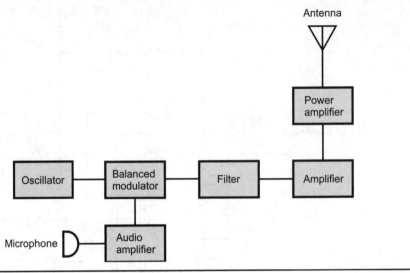

Figure 1-12   Block diagram of a basic SSB transmitter.

operate in a *linear* manner to prevent distortion and unnecessary spreading of the signal bandwidth, a condition that some engineers and radio operators call *splatter*. The term *linear* refers to the fact that the amplifier output varies in direct proportion to the input, so that if you graph the output as a function of the input, you get a straight line.

Frequency Modulation

In *frequency modulation* (FM), the instantaneous signal strength remains constant, and the instantaneous frequency varies. In the olden days of radio (and sometimes still today), engineers obtained FM by applying an audio signal to a *varactor diode* in a *voltage-controlled oscillator* (VCO). Figure 1-13 shows an example of this scheme, known as *reactance modulation*. The fluctuating voltage across the varactor causes its capacitance to change in accordance with the audio waveform. The fluctuating capacitance causes variations of the resonant frequency of the *inductance-capacitance* (*LC*) tuned circuit, causing small variations in the frequency generated by the oscillator. In the scenario of Fig. 1-13, the oscillator itself would be a so-called *Colpitts* circuit, which uses a split capacitance and a simple inductance.

Phase Modulation

You can indirectly obtain FM if you modulate the *phase* of an oscillator signal. That means you move the wave back and forth in time, without changing the frequency directly. However, when you vary the phase of a wave rapidly, you'll get variations in the instantaneous frequency as an indirect consequence. As a matter of fact, any instantaneous phase change shows up as an instantaneous frequency change and vice versa. But there's a glitch: When you employ so-called *phase modulation* (PM) with the intent of getting an FM signal, you must process the audio before you apply it to the modulator, adjusting the *frequency response* of the audio amplifiers by experiment. Otherwise the signal will sound distorted when you listen to it in a receiver designed for ordinary FM. Phase modulation is employed in many modern communications radios because it lends itself to the *frequency synthesizers* that have replaced tuned LC variable-frequency oscillators.

Tech Tip

Phase modulation works all right for voice communications, but some engineers think that it does not work as well as direct FM for high-fidelity applications, such as music broadcasting. These people say that the best audio fidelity in an FM radio receiver comes only as a result of direct FM using reactance modulation, such as the scheme shown in Fig. 1-13.

Deviation in FM and PM

In an FM or PM signal, the *deviation* is the maximum extent to which the modulated-carrier frequency rises above and falls below the unmodulated-carrier frequency. For most FM and PM voice transmitters, the deviation is standardized at ±5 kHz, a mode called *narrowband FM* (NBFM).

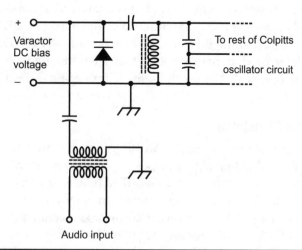

FIGURE 1-13   Generation of a frequency-modulated (FM) signal by means of reactance modulation.

In NBFM, the signal bandwidth is roughly the same as that of an AM signal containing the same modulating information (your voice, for example). In hi-fi music broadcasting, and in some other applications, such as high-speed data transmission, the deviation exceeds ±5 kHz, a mode called *wideband FM* (WBFM).

The deviation obtainable with FM is greater for a given oscillator frequency than the deviation that you get with PM. However, you can increase the deviation of any FM or PM signal with a *frequency multiplier*. When the signal passes through a frequency multiplier, the deviation gets multiplied along with the carrier frequency itself.

For Techies Only

In any FM or PM transmission, the deviation in the final output should equal or exceed the highest modulating audio frequency if you want to get optimum performance. Therefore, ±5 kHz is more than enough deviation for voice communications. For music, a deviation of ±15 kHz or ±20 kHz is required for good sound quality when the signal finally reaches people's receivers.

Pulse-Amplitude Modulation

You can modulate a signal by varying some aspect of a sequence, or *train*, of signal bursts called *pulses*. In *pulse-amplitude modulation* (PAM), the amplitude (strength) of each individual pulse varies according to the modulating waveform. Figure 1-14A shows an amplitude-versus-time graph of a hypothetical PAM signal. The modulating waveform appears as a dashed curve, and the pulses appear as vertical gray bars. Normally, the pulse amplitude increases as the instantaneous modulating-signal level increases (*positive PAM*). But you can reverse the situation so that higher audio levels cause the pulse amplitude to go down (*negative PAM*). Then the signal pulses are at their strongest when there is no modulation. The transmitter has to work harder to produce negative PAM than it does to produce positive PAM. Either way, all the pulses all last for the same length of time. That is to say, they're all equally wide.

Pulse-Width Modulation

You can modulate the output of an RF transmitter by varying the duration, or width, of the individual signal pulses to obtain *pulse-duration modulation* (PDM), also known as *pulse-width modulation* (PWM), as shown in Fig. 1-14B. Normally, the pulse width increases as the instantaneous modulating-signal level increases (*positive PWM*). But this situation can be reversed (*negative PWM*). The transmitter must work harder to accomplish negative PWM than to produce positive PWM because it has to transmit a signal for a larger proportion of the time on the average. Either way, the peak pulse amplitude remains constant; they're all equally strong.

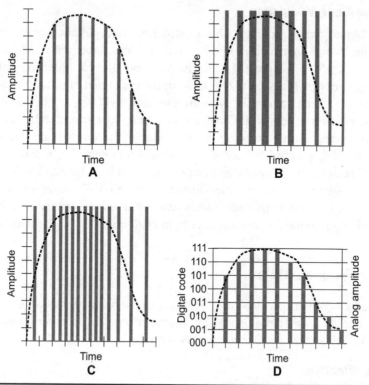

FIGURE 1-14   Time-domain graphs of various modes of pulse modulation. At A, pulse-amplitude modulation (PAM); at B, pulse-width modulation (PWM), also called pulse-duration modulation (PDM); at C, pulse-interval modulation (PIM); at D, pulse-code modulation (PCM).

Pulse-Interval Modulation

Even if all the pulses have the same amplitude and the same duration, you can obtain pulse modulation by varying how often the pulses occur. In PAM and PWM, you always transmit the pulses at the same time interval, known as the *sampling interval*. But in *pulse-interval modulation* (PIM), pulses can occur more or less frequently than they do under conditions of no modulation. Figure 1-14C shows a hypothetical PIM signal. Every pulse has the same amplitude and the same duration, but the time interval between them changes. When the transmitted signal has no modulation, the pulses emerge evenly spaced with respect to time. An increase in the instantaneous data amplitude might cause pulses to be sent more often, as is the case in the illustration here (*positive PIM*). Alternatively, an increase in instantaneous data level might slow down the rate at which the pulses emerge (*negative PIM*). As with PWM, a transmitter must work harder to generate negative PIM than to produce positive PIM.

Pulse-Code Modulation

In *digital communications*, the modulating data attains only certain defined states, rather than continuously varying. Compared with old-fashioned *analog communications* (where the state is continuously variable), digital modes offer improved *signal-to-noise* (S/N) *ratio*, narrower signal bandwidth, better accuracy, and enhanced reliability. In *pulse-code modulation* (PCM), any of the above-described aspects—amplitude, width, or interval— of a pulse train can be varied. But rather than having infinitely many possible states, the number of states equals some power of 2, such as 2^2 (four states), 2^3 (eight states), 2^4 (16 states), 2^5 (32 states), 2^6 (64 states), and so on. As you increase the number of states, the fidelity improves, but the signal gets more "complicated." Figure 1-14D shows an example of eight-level PCM. The amplitude levels are rendered as binary numbers from 000 (equivalent to the decimal number 0) to 111 (equivalent to the decimal number 7).

Receiver Fundamentals

A radio receiver converts radio waves into the original messages sent by a distant transmitter. Let's define a few important criteria for receiver operation, and then we'll look at a couple of common receiver designs.

Specifications

The *specifications* of a receiver quantify how well the hardware can carry out the tasks that its engineers designed and built it to do. Here are a few of the ones that you should think about before you buy a shortwave radio.

   Sensitivity—The most common way to express receiver sensitivity is to state the number of signal *microvolts* (millionths of a volt) that must exist at the antenna terminals to produce a certain *signal-to-noise ratio* (S/N) or *signal-plus-noise-to-noise ratio* (S+N/N) in relative amplitude units called *decibels* (dB). The sensitivity depends on the gain of the *front end* (the amplifier that the signal enters from the antenna). The amount of noise that the front end generates also matters because subsequent stages amplify its noise output as well as its signal output.
   Selectivity—The *passband*, or range of frequencies that the receiver can "hear," is established by a wideband *preselector* in the early RF amplification stages, and is honed to precision by *narrowband filters* in later amplifier stages. A typical preselector makes the receiver most sensitive within a few percentage points of the desired signal frequency. The narrowband filter responds only to the frequency or channel of a specific signal that you want to hear, and rejects signals in nearby channels.
   Dynamic range—The signals at a receiver input can vary over several orders of magnitude (powers of 10) in terms of absolute voltage. Engineers define

dynamic range as the ability of a receiver to maintain a fairly constant output, and still keep its rated sensitivity, in the presence of signals ranging from extremely weak to extremely strong. A good receiver exhibits dynamic range in excess of 100 dB, which translates to a signal input power ratio of 10,000,000,000 to 1! A competent technician can conduct experiments to determine the dynamic range of any receiver.

Noise figure—As the amount of internal noise a receiver produces goes down, the S/N ratio improves, if all other factors remain constant. You can expect an excellent S/N ratio in the presence of weak signals only when your receiver has a low noise figure, which is a measure of internally generated circuit noise. The noise figure makes the most practical difference at VHF, UHF, and microwave frequencies. *Gallium-arsenide field-effect transistors* (GaAsFETs) are known for the low levels of electrical noise they generate at these frequencies. You can get away with other types of FETs at lower frequencies. Bipolar transistors, which carry higher currents than FETs, generate more circuit noise than FETs do.

Direct-Conversion Receiver

A *direct-conversion receiver* derives its output by mixing incoming signals with the output of a tunable (variable frequency) *local oscillator* (LO). The received signal goes into a circuit called a *mixer* along with the output of the LO, and the two signals beat against each other in the mixer. Figure 1-15 is a block diagram of a direct-conversion receiver.

For reception of on/off keyed Morse code, the LO, which techies called a *beat-frequency oscillator* (BFO) before about 1980, is set a few hundred hertz above or below the signal frequency. You can also use this scheme to receive FSK signals. The audio output frequency equals the difference between the LO frequency and the incoming carrier frequency. For reception of AM or SSB signals, you must adjust the LO to precisely the same frequency as that of the signal carrier, a condition called *zero beat* because the *beat frequency*, or difference frequency, between the LO and the signal carrier is zero.

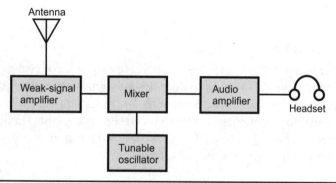

Figure 1-15   Block diagram of a direct-conversion receiver.

A direct-conversion receiver provides relatively poor *selectivity*, meaning that it can't do a good job of separating signals when they lie close together in frequency. In a direct-conversion receiver, you can hear signals on either side of the LO frequency at the same time, and you can't tell which one lies above the LO frequency and which one lies below it unless you adjust the tuning dial. Then one signal goes up in pitch and the other goes down, a weird and rather maddening effect to some folks. A *selective filter* can theoretically eliminate this problem. Such a filter must be designed for a fixed frequency if you expect it to work well. However, in a direct-conversion receiver, the RF amplifier must operate over a wide range of frequencies, making effective filter design an engineering challenge.

Superheterodyne Receiver

A *superheterodyne receiver*, also called a *superhet*, uses one or more LOs along with *signal mixers* to obtain a constant-frequency signal. You can more easily filter a fixed-frequency signal than you can filter a signal that changes in frequency (as it does in a direct-conversion receiver). A mixer produces output signals at the sum and difference frequencies of the input signals.

In a superhet, the incoming signal goes from the antenna through a tunable, sensitive front end, which comprises a precision weak-signal amplifier. The output of the front end mixes (heterodynes) with the signal from a tunable, unmodulated LO. You can choose the sum signal or the difference signal for subsequent amplification. Engineers call this signal the *first intermediate frequency* (IF), which can be filtered to obtain enhanced selectivity.

If the first IF signal passes straight into the *detector* (signal demodulator), you have a *single-conversion receiver*. Some receivers use a second mixer and second LO, converting the first IF to a lower-frequency *second IF*. Then you have a *double-conversion receiver*. The IF bandpass filter can be constructed for use on a fixed frequency, allowing superior selectivity and facilitating adjustable bandwidth. The sensitivity is enhanced because fixed-frequency IF amplifiers are easy to keep in tune. In a double-conversion receiver, the low second IF makes it possible to obtain better selectivity than can normally be obtained with a single-conversion design.

For Techies Only

Unfortunately, even the best superheterodyne receiver can intercept or generate unwanted signals. External false signals are called *images*; internally generated false signals are known as *birdies*. If you carefully choose the LO frequency (or frequencies) when you design a superhet, images and birdies will rarely cause problems during ordinary operation. But you'll always find some if you look hard enough for them!

Stages of a Single-Conversion Superhet

Figure 1-16 shows a block diagram of a single-conversion superheterodyne receiver. Individual receiver designs vary, but you can consider this example representative. The stages break down as follows:

- The front end serves as the first RF amplifier, and may include a *bandpass filter* between the amplifier and the antenna. The dynamic range and sensitivity of a receiver are determined largely by the performance of the front end.
- The mixer stage, in conjunction with the tunable LO, converts the variable signal frequency to a constant IF. The output occurs at either the sum or the difference of the signal frequency and the LO frequency.
- The IF stages produce most of the gain (amplification) in a superhet receiver. You also get most of the selectivity here, filtering out unwanted signals and noise while allowing the desired signal to pass.
- The detector extracts the information from the signal. Common circuits include the *envelope detector* for AM, the *product detector* for SSB, FSK, and CW, and the *ratio detector* for FM and PM. You'll find explanations of these system elements later in this chapter.
- The audio amplifier boosts the demodulated signal to a level suitable for a speaker or headset. Alternatively, you can feed the signal to a printer, facsimile machine, computer, or other device.

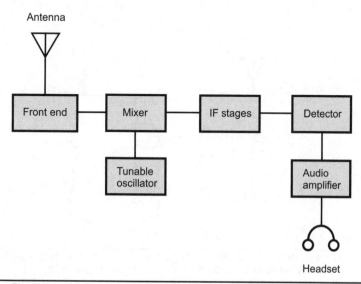

FIGURE 1-16   Block diagram of a single-conversion superheterodyne receiver.

Predetector Stages

When you design and build a superheterodyne receiver, you must ensure that the stages preceding the first mixer provide reasonable gain but generate minimal internal noise. They must also be able to bring in strong signals without *desensitization* (losing gain), a phenomenon also known as *overloading*.

Preamplifier

If a receiver doesn't have enough sensitivity to satisfy you, you can use a *preamplifier* between the antenna and the radio. Figure 1-17 shows a simple preamplifier circuit that can use either a *junction field-effect transistor* (JFET) or a GaAsFET. Input tuning reduces noise and provides some selectivity. This circuit produces 5 dB to 10 dB of useful signal gain, depending on the frequency and the choice of FET.

You must make sure that your preamplifier remains linear in the presence of strong input signals. Nonlinearity can cause unwanted mixing among multiple incoming signals. These so-called *mixing products* produce *intermodulation distortion* (IMD) or *intermod* that can spawn multiple images inside the receiver. Intermod can also degrade the S/N ratio by generating *hash*, a form of wideband RF noise.

Confused?

If you aren't familiar with schematic diagrams like the one in Fig. 1-17, why not learn how to read and draw them right now? If you get serious about hobby radio, and especially Amateur Radio, you'll want to get comfortable with these diagrams. A good tutorial for this purpose is *Beginner's Guide to Reading Schematics*, 3rd edition (McGraw-Hill, 2013).

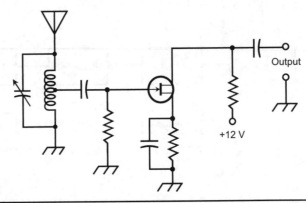

FIGURE 1-17   A tunable preamplifier. Normally, this type of amplifier uses a JFET or GaAsFET, and goes between the antenna and the front end of the radio.

Front End

At low and medium frequencies, considerable atmospheric noise exists, and the design of a front-end circuit is simple because you don't have to worry very much about internally generated noise. Conditions are bad enough in the antenna feed line as it comes into your receiver; a little more noise coming from inside the radio itself won't likely matter.

Atmospheric noise diminishes as you get above 30 MHz or so. Then the main sensitivity-limiting factor becomes noise generated within the receiver. For this reason, front-end design grows in importance as the frequency rises through the VHF, UHF, and microwave parts of the radio spectrum.

The front end, like a preamplifier, must remain as linear as possible. In other words, it mustn't introduce distortion. The greater the degree of nonlinearity, the more susceptible the circuit becomes to the generation of mixing products and intermod. The front end should also have the greatest possible dynamic range.

Preselector

A preselector, if the radio has one, provides a bandpass response that improves the S/N ratio, and reduces the likelihood of overloading by a strong signal that's far removed from the operating frequency. The preselector also provides some extra *image rejection* in a superheterodyne circuit. You can tune a preselector by means of *tracking* with the receiver's main tuning control, but this technique requires careful design and alignment. Some older receivers incorporate preselectors that must be adjusted independently of the receiver tuning.

IF Chains

A high IF (at least several megahertz) works better than a low IF for image rejection. However, a low IF allows for superior selectivity. Double-conversion receivers use two mixers and two IF chains. The comparatively high *first IF* and low *second IF* give you the "best of both worlds." The designers of these receivers *cascade* multiple IF amplifiers (connect them one after the other). There are two sets, called *chains*, of IF amplifiers. The *first IF chain* follows the first mixer and precedes the second mixer, and the *second IF chain* follows the second mixer and precedes the detector.

Engineers sometimes express IF-chain selectivity by comparing the bandwidths for two power-attenuation (amplitude-reduction) values, usually −3 dB and −30 dB, also called *3 dB down* and *30 dB down*. This specification offers a good description of the bandpass response. You call the ratio of the bandwidth at −30 dB to the bandwidth at −3 dB the *shape factor*. In general, small shape factors are more desirable than large ones, but small shape factors are more difficult to attain in practice.

Did You Know?

When you have a receiver with a small shape factor and then graph the system gain as a function of frequency, you get a curve that has a flat top and steep sides. Because it somewhat resembles a rectangle, engineers say that the receiver has a *rectangular response*.

Detectors

Detection, also called *demodulation*, allows a radio receiver to recover the modulating information, such as audio, images, or printed data, from an incoming signal.

Detection of AM

A radio receiver can extract the information from an AM signal by "chopping off" either the positive or the negative part of the carrier wave (in other words, rectifying it), and then filtering the output waveform just enough to smooth out the RF pulsations. Figure 1-18A shown a simplified time-domain view of how this process works. The rapid pulsations (solid curves) occur at the RF carrier frequency; the slower fluctuation (dashed curve) portrays the modulating data. The carrier pulsations get smoothed out as the output passes through a capacitor large enough to hold the charge for one carrier current cycle, but not so large that it dampens or obliterates the fluctuations in the modulating signal. Engineers call this technique *envelope detection*.

Detection of CW and FSK

If you want a receiver to detect CW signals, you must inject a constant-frequency, unmodulated carrier a few hundred hertz above or below the signal frequency. The local carrier is produced by a tunable LO inside the receiver. The LO signal and the incoming CW signal beat against each other in a mixer to produce audio output at the sum or difference frequency. You can tune the LO to obtain an audio note at a comfortable listening pitch, usually 500 to 1000 Hz. This process is called *heterodyne detection*.

You can detect FSK signals using the same method as CW detection. The carrier beats against the LO in the mixer, producing an audio tone that alternates between two different pitches. With FSK, the LO frequency is set a few hundred hertz above or below both the *mark frequency* and the *space frequency*. The *frequency offset*, or difference between the LO and signal frequencies, determines the audio output frequencies. You adjust the frequency offset to get specific standard AF notes (such as 2125 Hz and 2295 Hz in the case of 170-Hz shift).

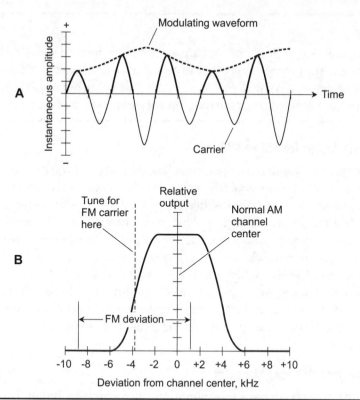

FIGURE 1-18A-B   At A, envelope detection of AM, shown in the time domain (a graph of amplitude versus time). At B, slope detection of FM, shown in the frequency domain (a graph of amplitude versus frequency).

Slope Detection of FM and PM

You can use an AM receiver to detect FM or PM by setting the receiver frequency near, but not exactly at, the unmodulated-carrier frequency. An AM receiver has a filter with a passband of a few kilohertz and a selectivity curve, such as the one shown in Fig. 1-18B. If you tune the receiver so that the FM unmodulated-carrier frequency lies near either edge, or *skirt*, of the filter response, frequency variations in the incoming signal cause its carrier to swing in and out of the receiver passband. As a result, the instantaneous receiver output amplitude varies along with the modulating data on the FM or PM signal. In this system, known as *slope detection*, a nonlinear relationship exists between the instantaneous deviation and the instantaneous output amplitude because the skirt of the passband does not produce a straight line in a graph of amplitude versus frequency.

> **Tip**
>
> Slope detection does not provide an optimum method of detecting FM or PM signals. The process, which causes audio distortion because of the nonlinearity in the response, can usually yield an intelligible voice, but it will ruin the quality of music, and may disrupt the reception of certain kinds of data signals.

Discriminator for FM or PM

A *discriminator* produces an output voltage that depends on the instantaneous signal frequency. When the signal frequency lies at the center of the receiver passband, the output voltage equals zero. When the instantaneous signal frequency falls below the passband center, the output voltage becomes positive. When the instantaneous signal frequency rises above center, the output voltage becomes negative.

In a discriminator circuit, you get a linear relationship between the instantaneous deviation of the FM (which can result indirectly from PM) and the instantaneous output amplitude. Therefore, the detector output represents a faithful reproduction of the incoming signal data. A discriminator will respond to amplitude variations, but you can use an amplitude limiting circuit to overcome this effect if it poses a problem.

Ratio Detector for FM or PM

A *ratio detector* comprises a discriminator with a built-in limiter. The original design was developed by RCA (Radio Corporation of America), and works well in hi-fi receivers and in the audio portions of old-fashioned analog TV receivers. Figure 1-18C illustrates a ratio detector circuit. The potentiometer marked "balance" must be adjusted experimentally to get optimum received-signal audio quality.

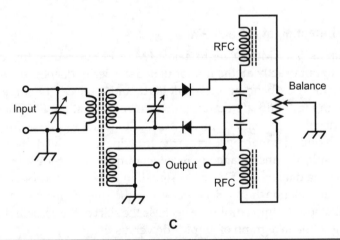

C

FIGURE 1-18C   At C, a ratio detector circuit for demodulating FM signals.

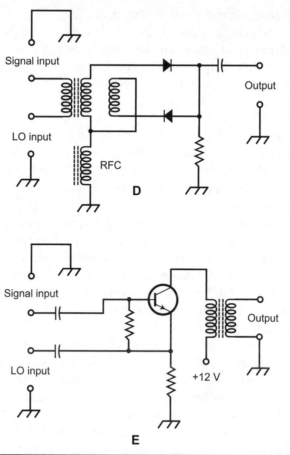

FIGURE 1-18D-E   At D, a product detector using diodes. At E, a product detector using an NPN bipolar transistor biased as a class-B amplifier.

Detection of SSB

For reception of SSB signals, most communications engineers favor a *product detector*, although a direct-conversion receiver can do the job. A product detector also facilitates reception of CW and FSK. The incoming signal combines in a mixer with the output of an unmodulated LO, reproducing the original modulating signal data. Product detection occurs at a single frequency, rather than at a variable frequency that is characteristic of direct-conversion reception.

Figures 1-18D and E show product-detector circuits, which can also serve as mixers in superheterodyne receivers. In the circuit at D, diodes are used. They're *passive components*, so you don't get any amplification. The circuit at E employs a single-transistor amplifier, providing some gain if the incoming signal has been sufficiently amplified by the front end before it arrives at the detector input.

The effectiveness of the circuits shown in Figs. 1-18D or E lies in the nonlinearity of the semiconductor devices. The nonlinearity facilitates (and, in fact, encourages) the heterodyning that's necessary to obtain sum and difference frequency signals that result in data output. So, as you can see, nonlinearity isn't always bad. Sometimes you *want* a circuit to be nonlinear!

Audio Filtering

In a communications system, a human voice signal requires a band of frequencies ranging from about 300 Hz to 3000 Hz for a listener to easily understand the content. An *audio bandpass filter*, with a passband of 300 Hz to 3000 Hz, can improve the intelligibility in some voice receivers. An ideal voice audio bandpass filter has little or no attenuation within the passband range but high attenuation outside the passband range, along with a near-rectangular response curve.

A CW or FSK signal requires only a few hundred hertz of bandwidth. Audio CW filters can narrow the response bandwidth to 100 Hz or less, but passbands narrower than about 100 Hz produce *ringing*, degrading the quality of reception at high data speeds. With FSK, the bandwidth of the filter must be at least as large as the difference (shift) between mark and space, but it need not (and shouldn't) greatly exceed the frequency shift.

An *audio notch filter* is a *band-rejection filter* with a sharp, narrow response. Band-rejection filters pass signals only below a certain lower cutoff frequency or above a certain upper cutoff frequency. Between those limits, in the so-called *bandstop range*, signals are blocked. A notch filter can mute an interfering unmodulated carrier or CW signal that produces a constant-frequency tone in the receiver output. Audio notch filters are tunable from at least 300 Hz to 3000 Hz. A good notch filter has an extremely narrow bandstop range, so you can get rid of an unwanted carrier signal with minimal adverse effect on the rest of the audio range.

Did You Know?

Some audio notch filters work automatically; when an interfering audio tone appears, the notch finds and mutes it within a few tenths of a second.

Squelching

A *squelch* silences a receiver when no incoming signals exist, allowing reception of signals when they appear. Most FM communications receivers use squelching systems. The squelch is normally *closed*, cutting off all audio output (especially receiver hiss, which annoys some communications operators) when no signal is present. The squelch *opens*, allowing everything to be heard, if the signal amplitude exceeds a *squelch threshold* that the operator can adjust.

In some radios, the squelch does not open unless an incoming signal has certain predetermined characteristics. This feature is called *selective squelching*. The most common way to achieve selective squelching is the use of a *subaudible* (below 300 Hz) *tone generator* or audio *tone-burst generator* in the transmitter. The squelch opens only in the presence of signals modulated by an audio tone, or sequence of tones, having the proper characteristics. Some radio operators use selective squelching to prevent unwanted transmissions from coming in.

Exotic Communications Methods

Communications engineers have a long history of innovation, developing alternative (and some downright weird) wireless modes. In recent years, new modes have emerged; you can expect more to come, each of which offers specific advantages under strange or difficult conditions. Four common examples follow.

Dual-Diversity Reception

A *dual-diversity receiver* can reduce fading in radio reception at frequencies between approximately 3 MHz and 30 MHz when signals propagate through the ionosphere and return to earth's surface. The system comprises two identical receivers tuned to the same signal and having separate antennas spaced several wavelengths apart. The outputs of the receiver detectors go into a single audio amplifier, as shown in Fig. 1-19.

Dual-diversity receiver tuning is a sophisticated technology (you might call it an art), and good equipment for this purpose costs a lot of money. Some advanced

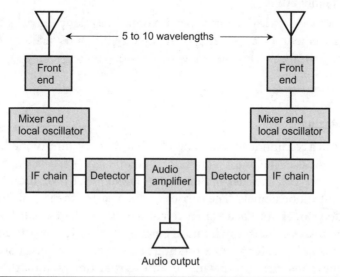

FIGURE 1-19   Block diagram of a diversity radio reception system.

diversity-reception installations employ three or more antennas and receivers, providing superior immunity to fading, but further compounding the tuning difficulty and driving up the expense.

Synchronized Communications

Digital signals require less bandwidth than analog signals to convey a given amount of information per unit of time. The term *synchronized communications* refers to any of several specialized digital modes in which the transmitter and receiver operate from a common frequency-and-time standard to optimize the amount of data that can be sent in a communications channel or band.

In synchronized digital communications, also called *coherent communiations*, the receiver and transmitter operate in lockstep. The receiver evaluates each transmitted data bit for a block of time lasting for the specified duration of a single bit. This process makes it possible to use a receiving filter having extremely narrow bandwidth. The synchronization requires the use of an external frequency-and-time standard, such as that provided by the *National Institute of Standards and Technology* (NIST) radio station WWV in the United States. Frequency dividers generate the necessary synchronizing signals from the frequency-standard signal. A tone or pulse appears in the receiver output for a particular bit if, but only if, the average signal voltage exceeds a certain value over the duration of that bit. False signals caused by filter ringing, atmospheric noise, or ignition noise are generally ignored, because they rarely produce sufficient average bit voltage.

For Techies Only

Experiments with synchronized communications have shown that the improvement in S/N ratio, compared with nonsynchronized systems, is several decibels at low to moderate data speeds.

Multiplexing

Signals in a communications channel or band can be intertwined, or *multiplexed*, in various ways. The most common methods are *frequency-division multiplexing* (FDM) and *time-division multiplexing* (TDM). In FDM, the channel is broken down into subchannels. The carrier frequencies of the signals are spaced so that they don't overlap. Each signal remains independent of all the others. A TDM system breaks signals down into segments of specific time duration, and then the segments are transferred in a rotating sequence. The receiver stays synchronized with the transmitter by means of an external time standard, such as the data from WWV.

For Techies Only

Multiplexing requires an *encoder* that combines or "intertwines" the signals in the transmitter, and a *decoder* that separates or "untangles" the signals in the receiver.

Spread Spectrum

In *spread-spectrum communications*, the transmitter varies the main carrier frequency in a controlled manner, independently of the signal modulation. The receiver is programmed to follow the transmitter frequency from instant to instant. The whole signal therefore roams up and down in frequency within a defined range.

In spread-spectrum mode, the probability of *catastrophic interference*, in which one strong unwanted signal can obliterate the desired signal, is near zero. Unauthorized people find it impossible to eavesdrop on a spread-spectrum communications link unless they gain access to the *sequencing code*, also known as the *frequency-spreading function*. Such a function can be complicated indeed. As long as neither the transmitting operator nor the receiving operator divulge the sequencing code to anyone else, then (ideally) no unauthorized listener can intercept it.

During a spread-spectrum contact between a given transmitter and receiver, the operating frequency can fluctuate over a range of several kilohertz, megahertz, or tens of megahertz. As a band becomes occupied with an increasing number of spread-spectrum signals, the overall noise level in the band appears to increase. Therefore, a practical limit exists to the number of spread-spectrum contacts that a band can handle. This limit is roughly the same as it would be if all the signals were constant in frequency, and had their own discrete channels. The main difference between fixed-frequency communications and spread-spectrum communications, when the band gets crowded, lies in the *nature* of the mutual interference.

A common method of generating spread-spectrum signals involves so-called *frequency hopping*. The transmitter has a list of channels that it follows in a certain order. The transmitter jumps from one frequency to another in the list. The receiver must be programmed with this same list, in the same order, and must be synchronized with the transmitter. The *dwell time* equals the length of time that the signal remains on any given frequency; it's the same as the time interval at which frequency changes occur. In a well-designed frequency-hopping system, the dwell time is short enough so that an unauthorized listener using a receiver set to a constant frequency won't notice it. In addition, the signal should not dwell on any frequency long enough to cause interference with a fixed signal that happens to lie on that frequency. The transmitter sequence contains numerous *dwell frequencies*, so the signal energy gets diluted to the extent that, if someone tunes to any particular frequency in the sequence, they won't notice the signal.

Another way to obtain spread-spectrum emission, called *frequency sweeping*, requires frequency-modulating the main transmitted carrier with a waveform that guides it smoothly up and down in frequency over the assigned band. The "sweeping FM" remains entirely independent of the actual data that the signal conveys. A receiver can intercept the signal if, but only if, its instantaneous frequency varies according to the same waveform, over the same band, at the same rate, and in the same phase as that of the transmitter. The transmitter and receiver in effect roam all over the band, following each other from moment to moment according to a "secret map" that only they know.

Did You Know?

However exotic it might seem at first thought, spread-spectrum technology has existed for quite a long while. Among the first people to dream it up were the composer George Antheil and the actress Hedy Lamarr, who received a U.S. patent for a spread-spectrum process in 1942!

CHAPTER 2

Shortwave and Allwave Listening

The frequency range from 3 MHz to 30 MHz (wavelengths of 100 meters to 10 meters) is sometimes called the *shortwave band*. You can build or buy a shortwave radio receiver, install an outdoor wire antenna, and listen to signals from all around the world in an activity called *shortwave listening* (SWL). If you extend the receiving range below 3 MHz down to the lowest frequencies you can get a radio to bring in, and if you can also extend the range above 30 MHz up to the highest frequencies you can manage, you can think of the hobby as *allwave listening* (AWL). In the United States, you don't need any sort of license to engage in radio listening at any frequency. In some countries, however, you will have to get a license to receive radio signals (believe it or not). If you don't live in the United States, check with your government authorities before you get involved with SWL or AWL.

How I Got into SWL

My SWL experience dates back to elementary school in the 1960s, when my dad bought me a radio project kit for my birthday. I was in the third grade, and even as I write this chapter, I have a note that I wrote to myself that year (1963), stating that I intended to become an Amateur Radio operator by the time I turned 10 years old. (Actually it took me until the sixth grade, at age 12.) I don't know how I would have heard about ham radio, were it not for SWL in those days of stuffing pieces of wire into spring contacts on a circuit board with capacitors, coils, resistors, and a 1T4 vacuum tube about the size of my thumb that ran off of a 45-volt battery and a 1.5-volt flashlight cell.

The first few projects involved simple broadcast receivers, such as a "crystal set" that worked without a battery, progressing to a battery-powered radio with an audio amplifier after the detector, and then adding other components to increase the selectivity and sensitivity. I recall hearing WLS in Chicago on that radio at night, and marveling at the vast distances over which radio waves could travel over free space to reach me, in my bedroom with that earphone clamped against my head in Rochester, Minnesota! My fascination with radio began back then.

Later projects with that kit included substitution of a different antenna coil, one that looked smaller than the original. It supposedly provided shortwave reception, but it didn't bring in much of anything until I got to the project called the "Regenerative Grid Leak Detector with Pentode Amplifier." Then I heard signals from all over the world: strange languages, bizarre buzzing and swishing and whining noises, and a time-standard station that claimed to come out of some observatory in Canada. I could tune through all of this cool stuff with a variable capacitor no larger than a golf ball.

An Old Timer Remembers (Not!)

I don't recall the component values or the schematic diagram for that little regenerative shortwave radio; the kit provided only pictorial layouts. It probably had a circuit configuration similar to the one shown in Fig. 2-1.

My dad introduced me to other hobbies aside from radio; he was the sort of father every kid should have, encouraging me to find and chase my own interests. He never told me where to "sail," but he made sure that I had good "boats." He

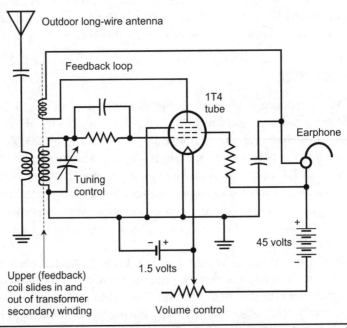

FIGURE 2-1   Generic diagram of the circuit for my first shortwave radio project, according to my best recollection. I don't remember the component values.

bought me a small refracting telescope to complement his own big Newtonian reflector; I got a pretty good microscope one year too. But it was that radio kit that pointed me toward my final career: to work for companies that dealt with Amateur Radio issues and equipment, and later to write about hobby electronics in books like the one you're reading right now.

Tip

Shortwave listening can help you find out whether or not you have enough interest in Amateur Radio to go ahead and get an operator's license! Learning the electronics theory and regulations well enough to pass the ham radio tests will take some work, but if your heart's in it, it will seem like play.

In the fifth grade, I got a more sophisticated, commercially manufactured shortwave radio receiver, the Hallicrafters model SX-130. I recall its appearance in every detail. It used a single-conversion, superhet design of the sort shown back in Chap. 1 (Fig. 1-16), along with extra audio amplifiers so that the sound could drive a loudspeaker. Like nearly all shortwave radios in the mid-1960s, the SX-130 worked with vacuum tubes. It covered frequencies from the bottom of the standard AM broadcast band at 535 kHz up to the top of that band at 1.605 MHz, had a little gap where it didn't receive anything, and then allowed reception over a continuous range of frequencies from 1.725 MHz to 31.500 MHz.

The SX-130, like most state-of-the-art shortwave radios of its day, employed two tuning dials, one called *main tuning* and the other called *bandspread*. The bandspread scale had a "slide-rule" geometry with multiple scales, but it gave accurate readings only when the main tuning dial was set at specific points. The SX-130 did not have an internal calibrator, so I had to guess at the exact frequency to which the radio was set. I learned how to "tweak" the main tuning dial to get reasonable bandspread scale accuracy by taking advantage of the National Bureau of Standards (now called the National Institute of Standards and Technology) time-and-frequency standard broadcasts. The station had the call sign WWV back then, as it does today. I would set the bandspread to a known WWV frequency and then carefully manipulate the main tuning dial to make the signal come in just right.

I sat for hours in front of that radio, fascinated, listening to the beeps and buzzes and tones and ticks that WWV produced (one of which I recognized as a sound effect in an episode of the original *Star Trek* television series). I learned that the Canadian station I had heard with the regenerative radio receiver kit had the call sign CHU, transmitting at 7.335 MHz. I discovered cryptic conversations taking place among individuals in frequency bands around 3.5 MHz, 7 MHz, and 14 MHz. I heard Morse code and other unidentifiable signals that sounded as if they could just as well have come from extraterrestrial civilizations. Before very long, I realized that these emissions came from hobby radio operators using transmitters in their own homes: radio hams!

When I entered the sixth grade in the fall of 1965, I started taking a ham radio class in my hometown, offered by the Rochester (Minnesota) Amateur Radio Club. In March of 1966, I got my Novice Class License, which, back in that day, offered severely restricted privileges. I had to learn to send and receive Morse code at five words per minute. I actually had to write down what I heard on the spot, for a solid continuous minute or more, without a single mistake. I had to send a solid minute of code error-free too, at that same five words per minute. But I finished the course somehow, and got my Novice Class License. My exuberance knew no bounds when I made my first contact on 7.185 MHz with a radio ham in Zeeland, Michigan, using Morse code that I pounded out with a simple telegraph key. I still have that key in my "ham shack" today, although I don't use it anymore.

Curious?

You can find photographs and specifications for the Hallicrafters SX-130 and similar vintage shortwave radios by conducting targeted Internet searches. For example, try entering "Hallicrafters SX-130" in the phrase box of an Internet search engine. You might attend a large Amateur Radio convention and find vintage radios on display there; you might even find one on sale. Some people make a hobby out of collecting and refurbishing old-time shortwave radios. Maybe you'll get into that activity!

Shortwave Broadcast Bands

International shortwave broadcasting takes place in several bands covering well-defined frequency intervals, with the lowest frequency (longest wavelength) at 2.300 MHz. Table 2-1 shows the shortwave broadcast bands designated at the World Radio Conference in 1997. The complete span of frequencies covers more than an order of magnitude (factor of 10), that is, the top frequency on 11 meters is a little more than 10 times the lowest frequency on 120 meters. Let's look at how these bands behave, so you'll have an idea of what to expect when you tune your radio to them.

Tech Note

As the frequency of a radio wave rises, the wavelength grows shorter. In other words, the frequency of a signal varies in *inverse proportion* to its wavelength. Figure 2-2 shows an example. Engineers and scientists define radio wavelengths for *free space* (the earth's atmosphere or the vacuum of outer space) unless they specify some other medium, such as a wire or cable.

TABLE 2-1   International Shortwave Radio Broadcast Bands

Frequency Range	Wavelength Designator
2.300 MHz to 2.495 MHz	120 meters
3.200 MHz to 3.400 MHz	90 meters
3.900 MHz to 4.000 MHz*	75 meters
4.750 MHz to 5.060 MHz	60 meters
5.900 MHz to 6.200 MHz	49 meters
7.200 MHz to 7.600 MHz*	41 meters
9.400 MHz to 9.900 MHz	31 meters
11.600 MHz to 12.200 MHz	25 meters
13.570 MHz to 13.870 MHz	22 meters
15.100 MHz to 15.800 MHz	19 meters
17.480 MHz to 17.900 MHz	16 meters
18.900 MHz to 19.020 MHz	15 meters
21.450 MHz to 21.850 MHz	13 meters
25.600 MHz to 26.100 MHz	11 meters

*Shared with Amateur Radio operators in some parts of the world.

120 Meters

Radio-wave propagation at 2.300 MHz resembles that at the upper end of the standard AM broadcast band. The effective range at night greatly exceeds the range during the day because when the sun is above the horizon, the ionosphere's D layer absorbs radio waves at this frequency before they can reach the higher layers. At night, the D layer disappears, and you can expect worldwide reception from stations

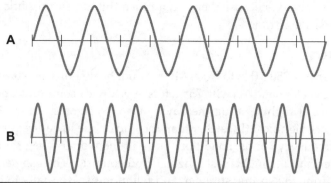

FIGURE 2-2   In free space, wavelength varies inversely with frequency. In this example, the wave at **A** has half the frequency of the wave at **B**. A complete wave cycle at **A**, therefore, measures twice the length of a complete wave cycle at **B**.

that use high RF output power levels (several kilowatts or more). Conditions peak during the winter months, especially at the higher latitudes where thunderstorms don't occur often at that time of the year. Thunderstorms produce *sferics* (sometimes called "static"), a form of *radio noise* that can impede or ruin signal reception. In addition, during the fall and winter, more of your hemisphere lies in darkness than is the case in the spring and summer, so you can enjoy the advantages of nighttime conditions for a greater portion of the day.

90 Meters

In the range of frequencies from 3.200 MHz to 3.400 MHz, signals propagate better at night than during the day, just as they do on the 120-meter band and the standard AM broadcast band. You might notice improved reception here, compared with 120 meters, from transmitting stations that use modest output power levels. If you have done much operating on the ham radio 80-meter and 75-meter bands (3.500 MHz to 4.000 MHz), you'll already have a good idea of how things work on the 90-meter shortwave broadcast band, and vice versa. As with 120 meters, conditions tend to peak in the winter (for the same reasons).

75 Meters

The shortwave band at 3.900 MHz to 4.000 MHz lies within the American ham radio 75-meter voice band, so ham radio operators must share the spectrum with these stations. Because shortwave broadcasting transmitters generally use more power than ham radios do, Amateur Radio operators sometimes find communication difficult in the top 100 kHz of their 75-meter band. When you listen to signals in this band, you can tell the difference between ham radio stations and shortwave broadcast stations. Ham radio operators use SSB emission, requiring a receiver with a local internal oscillator, but broadcast stations use AM emission, and you can bring them in with a simple envelope detector. Excellent worldwide shortwave reception can be had at night on this band, especially in the winter at latitudes far from the equator, because there's not much thunderstorm static and the darkness hours last a long time.

60 Meters

In the range 4.750 MHz to 5.060 MHz, you'll hear shortwave broadcasts on the so-called 60-meter band. As with 75 meters, worldwide communications and reception take place at night. During the daylight hours, signals can travel a few hundred kilometers, significantly farther than they do on the lower frequencies. As with the lower-frequency bands, conditions peak in the winter when thunderstorms occur less often and the nights last longer. Good conditions can sometimes exist on this band during the spring, summer, and fall months, especially in places such as the West Coast of the United States where thunderstorms rarely occur at any time of year. Ham radio operators have a small frequency allocation near this band. If you hear SSB or CW signals in this part of the radio spectrum, they're probably coming from ham radio stations, especially if they're around 5.3 MHz to 5.4 MHz.

Skip Tips

Engineers define *skip* as the tendency for signals to pass over certain geographical areas when they propagate by means of the ionosphere. At high frequencies, skip occurs often, especially above 5 MHz or so. Figure 2-3 shows the effect. A transmitting station X comes in at receiving station Z, located thousands of kilometers distant, but not at receiving station Y, even though Y lies closer to X than Z does. The F layer lacks the ionization density necessary to bend the signals back to the surface at a sharp enough angle to allow reception by Y, but the shallower angle for Z works out just fine. The region that the signals pass over is called the *skip zone*. It has a roughly circular shape, centered at the transmitting station's surface location, and it grows in radius as the frequency increases. At the lowest shortwave frequencies, you'll rarely observe a skip zone. At the highest frequencies, the zone can sometimes grow to an "infinite" radius, in the sense that signals never return to the earth at any distance. When a skip zone exists, it tends to expand during the hours of darkness and shrink during the hours of daylight.

49 Meters

In the 49-meter band (5.900 MHz to 6.200 MHz), conditions are much the same as they are at 60 meters, although you'll occasionally see a skip zone form. Also, atmospheric noise tends to present less of a problem, so you can get good reception

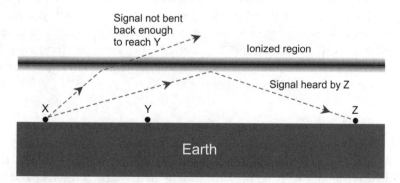

FIGURE 2-3   Station X transmits a signal that reaches the ionosphere, which bends back the waves enough to reach station Z but not enough to reach station Y. Station Y, therefore, lies within the skip zone with respect to station X at the frequency of interest.

during the spring, summer, and fall as well as in the winter. Once in a while, you'll hear stations as far away as 1000 kilometers during the daytime, but this band really "shines" at night, when you can expect to hear stations from all around the world. On a quiet winter evening, you'll derive a lot of enjoyment from tuning your radio to this band and "drilling down" to find the most interesting stations.

41 Meters

If you're a ham radio operator, you have a good idea of what conditions are like on the shortwave broadcast band that goes from 7.200 MHz to 7.600 MHz. If you prefer to use SSB on the 40-meter ham band, you've doubtless heard some of those powerful broadcast stations in the span where the bands overlap (7.200 MHz to 7.300 MHz). With your receiver's product detector switched on, shortwave broadcast stations define themselves as a loud carrier with "gibberish" superimposed. (When I first got my ham radio license in the 1960s, this problem was much worse than it is today, and sometimes 40 meters was practically unusable.) Daytime communication and reception can normally be had over distances up to about 1600 kilometers. At night, you'll hear signals from all over the world and at all times of the year. This band gets a lot of use; broadcasters like it because it practically guarantees a large listening audience.

Curious?

Do you wonder where the radio wavelength designators come from, and how accurate they really are? Here's a way to calculate the free-space wavelength of a radio wave if you know its frequency: Divide 300 (the approximate speed of light in millions of meters per second) by the frequency in megahertz. When you do that for the so-called "41-meter" shortwave broadcast band, you get a wavelength range of 39.5 meters at 7.600 MHz up to 41.7 meters at 7.200 MHz. The ham radio band known as "40 meters" extends from an actual wavelength of 41.1 meters at 7.300 MHz up to 42.9 meters at 7.000 MHz. If you want to claim technical accuracy based on the center frequencies of each band, you should probably call the ham band "42 meters" and leave the shortwave broadcast band at "41 meters." But alas: Once names get assigned in this branch of science and engineering, they usually stay that way.

31 Meters

Covering the half-megahertz range of 9.400 MHz to 9.900 MHz, the 31-meter shortwave broadcast band lies near the Amateur Radio 30-meter band (10.100 MHz to 10.150 MHz). If you've done much "hamming" on 30 meters, you'll know what to expect on the 31 meters and vice versa. Worldwide propagation can occur 24 hours a day here, although conditions are usually best over paths that lie entirely on the dark side of the planet. This band lies in a sort of "transition zone" as you go up in frequency, where the propagation begins to get better in the daytime than at night. The exact frequency transition point varies depending on the time of year. It also varies because of another phenomenon: the number of *sunspots* that fleck the face of our parent star. Sunspot numbers, on the average, increase and decrease in a rather well-defined cycle that lasts a little less than 11 years. You'll read more about this cycle shortly.

25 Meters

The shortwave broadcast band at 11.600 MHz to 12.200 MHz lies roughly half the way, in terms of frequency and wavelength, between the ham radio 30-meter and 20-meter bands. On 25 meters, you'll notice a big improvement, compared with the lower-frequency bands, in reception over paths that lie entirely on the sunlit side of the earth. You'll also notice that in the summertime, thunderstorms don't produce as much "static" at these frequencies as they do on the lower frequencies. Propagation can take place over the dark side of the earth, but sometimes the F layers are not sufficiently ionized at night to return signals. You can think of this effect as the "skip zone expanding to infinity." It's most likely to happen during times when the sunspot cycle is at, or near, its minimum.

22 Meters

Those of you familiar with the 20-meter ham band at 14.000 MHz to 14.300 MHz will know what conditions to expect on the shortwave broadcast band at 13.570 MHz to 13.870 MHz. And, for you SWL lovers who have thought about getting a ham radio license but keep putting it off, this band ought to give you a good feeling for what you can expect when you finally do get on the air and operate on 20 meters, the most popular of the ham bands! Worldwide reception occurs on the daylight side of the planet almost all the time, and when the sunspot cycle reaches its peak, good reception and communication can be had at night, as well. When the pioneers of shortwave communication got "all the way down" to wavelengths in this range, they learned, in full, how amazing the HF range of wavelengths can get. Some radio hams actually talked to each other from opposite sides of the globe using transmitters running only a few watts of RF output, along with modest wire antennas!

Heads Up!

Some people call long-distance ionospheric propagation "skip." That's a technically inaccurate use of the term, but you'll hear people make this mistake a lot, saying that they're "working skip" or "hearing skip." In its true sense, as you now know, the term *skip* refers to a condition or zone for which communication or reception *fails*.

19 Meters

Conditions in the frequency range from 15.100 MHz to 15.800 MHz resemble those on the 20-meter ham band and the 22-meter shortwave band. You might notice a slight improvement in daytime propagation, and a slight deterioration at night, compared with those other bands; in addition, conditions in the spring and summer are usually better than those in the fall and winter. This band covers a considerable frequency span, and you can expect to hear a lot of interesting stations. If you're listening on this band and you tune your radio down to 15.000 MHz, you'll often hear the time-and-frequency standard stations WWV in Colorado and/or WWVH in Hawaii. When I was a newly licensed ham radio operator in the 1960s, I took advantage of these stations' broadcasts to calibrate my radio for operation on the 20-meter ham band, and also to set my 24-hour clock, which ran off of standard utility electricity. Nowadays, digital radios offer near-perfect frequency calibration, and you can get a clock that automatically synchronizes itself to another time-and-frequency standard station, WWVL at 60 kHz, or else simply look at your Internet-connected computer's clock! So you probably won't need to use these stations to calibrate anything. (They sound cool on big surround-sound speakers, though, of the sort I have in my "shack"!)

16 Meters

The frequency range of 17.480 MHz to 17.900 MHz hosts international shortwave broadcasting in a band known as 16 meters. (Actually the wavelength comes closer to 17 meters.) Conditions here resemble those on the ham band called 17 meters at 18.068 MHz to 18.168 MHz, and vice versa. Worldwide communication and reception take place over paths that lie entirely on the daylight half of the earth nearly every day of every year. During times when the sunspot activity increases, conditions improve on this band, and paths will sometimes work even when they lie mostly or entirely on the dark side of the planet.

Here's a Twist

Once in awhile you'll hear signals that follow two paths, one signal going along a *geodesic arc* (great circle route) the short way around the earth, and the other signal going along a geodesic arc the long way around. This type of propagation can produce "reverb" on signals if the two paths differ in length (which they will, unless you happen to live exactly at the *antipode*, or opposite point on the earth, from the transmitting station). The reverberation, which sounds like the echo you hear when you shout in a room with bad acoustics, results from the fact that the *long-path* signal arrives a fraction of a second later than the *short-path* signal. This phenomenon will most likely show up at frequencies ranging from about 12 MHz to 18 MHz during times of peak sunspot activity.

For Nerds Only

If you want to know the actual wavelength ranges of all the shortwave broadcast bands, you can use the formula that I quoted above on each one of them, using Table 2-1 as your reference, and find out! For those of you who get a rush from exactitude, you can substitute 299.792 in place of 300 in the formula. (The speed of light in free space is 299,792 kilometers per second.)

15 Meters

Shortwave broadcasters refer to the range of frequencies from 18.900 MHz to 19.020 MHz as the 15-meter band, even though ham radio operators call their band at 21.000 MHz to 21.450 MHz by the same wavelength name. Actually the wavelengths at the middles of the bands are 15.8 and 14.1 meters, respectively. Conditions on these frequencies resemble those on the 16-meter band that lies about 1 MHz lower down in the frequency spectrum. The 15-meter shortwave broadcast band doesn't get much use. I checked it out on an afternoon with excellent reception taking place on the 16-meter and 13-meter shortwave bands as well as on the 15-meter ham band. I heard nothing between 18.900 MHz and 19.020 MHz.

13 Meters

The shortwave broadcast 13-meter band lies immediately above the ham radio 15-meter band in terms of frequency, covering a range of 21.450 MHz to 21.850 MHz. Conditions here are essentially the same as they are on the ham radio band at 21.000 MHz to 21.450 MHz. When the sun is in a "vibrant mood," you can expect good reception on this band for any path that exists entirely, or mostly, on the daylight side of the earth. For darkness paths, the situation gets more variable. During times of lackluster solar activity, this band often goes completely dead, with the skip zone in effect "expanding to infinity" 24 hours a day. Nevertheless, this band is worth checking out at any time. You never know what surprises it might bring!

Where's the Daylight Side?

If you want to know which part of the earth lies in sunlight and which part lies in darkness at any particular time, do an Internet search on the phrase "Grey Line Map" (note the British spelling). Some of the hits will lead you to maps that portray the contour of the *grey line* where the sun lies on the horizon as seen from the earth's surface. Along with the discussions in this chapter, any one of these maps can help you find out whether or not you should expect good shortwave propagation from any given transmitting location to any given receiving location. As of this writing, I found a good grey-line map, which refreshes itself every five minutes if you keep your computer connected to the Internet, at

dx.qsl.net/propagation/greyline.html

Remember that shortwave radio signals usually follow the shorter of two geodesic arcs around the planet as they propagate by means of the ionosphere. A geodesic will nearly always show up as a curve on an ordinary map (a good example is the grey line itself, which is always a great circle), and you might have a hard time guessing the geodesic's position between your location and the place where the transmitted signal originates. If you have access to a globe, you can use a piece of string to find the short and long geodesics between any points of your choice. You can also do an Internet search on the phrase "great circle map" and take it from there. Find a map that's centered near you. On a great circle map, any geodesic from the location at the center shows up as a straight, radial line that runs out from the center to the location in question.

11 Meters

Evidently this band, which goes from 25.600 MHz to 26.100 MHz, gets little or no use for the purpose of AM shortwave broadcasting. On an afternoon when signals were abundant on the 27-MHz Citizens Band (CB) and also on the ham radio 10-meter band, I heard nothing here other than a few strange-sounding "blips." If you've done much operating on the ham radio 12-meter band at 24.890 MHz to 24.990 MHz, or if you happen to be one of those CB "freebanders" who spends a lot of time on the air around 27 MHz, you have a good idea of what to expect in terms of propagation conditions on the 11-meter shortwave band. If anyone ever cares to take advantage of this band, worldwide coverage will often be possible over daylight paths during times of high solar activity. At night, or during solar minima, this band can go "dead" for weeks or months at a time, except for local communications up to a few kilometers.

How Do They Look?

Figure 2-4 shows the locations of the shortwave broadcast bands according to frequency. You can see that, when you combine their spectrum space, they take up a significant part of the HF radio zone. Remember that as the frequency goes up, the free-space wavelength gets shorter. While the entire set of bands covers more than an order of magnitude frequency-wise or wavelength-wise, each individual shortwave broadcast band is narrow, so you can expect that propagation conditions at the top of any given band will closely resemble conditions at the bottom of the same band.

Other Shortwave Activity

Of course, you'll find plenty of other activity on the shortwave frequencies besides international broadcasting and ham radio. The NIST broadcasts time and frequency standard signals in the United States from two locations, one in Colorado (WWV)

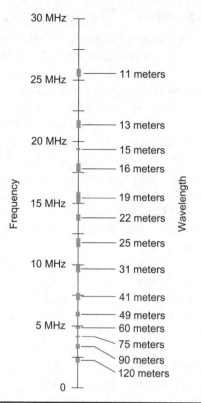

FIGURE 2-4   Nomograph showing the shortwave broadcast bands on a linear scale according to frequency. The frequency increases, and the wavelength gets shorter, as you move upward along the scale.

and the other in Hawaii (WWVH). I've heard these signals at 5, 10, and 15 MHz quite regularly. Go to the website

www.nist.gov

and do a search on "WWV" as a single word. You'll find plenty of information about their broadcast format on that site. They do a lot more than provide frequency standards and tell you the time!

As you explore the shortwave realm, you'll hear strange signals of all sorts: hissing and whining and roaring signals, mostly high-speed digital transmissions along with some jamming signals (yes, that stuff still happens) where one nation doesn't like what another nation has to say. Many of the signals have a military origin. If you happen to know the Morse code and can "copy" it at a reasonable speed, you might make sense out of some of these signals. My advice here is simple: Get a good general coverage receiver, set up a good antenna as described later in this chapter, and start listening!

A Simple Antenna for SWL

If you buy a commercially manufactured shortwave radio, especially the portable type that seems to prevail these days, it will likely have its own antenna attached: a whip that extends to about one meter in length. That antenna will work okay for receiving strong signals, but your best option for serious SWL is an outdoor *random wire* at least 20 meters long, and much longer if you can manage it.

Use insulated wire. Plain old lamp cord, also called "zip cord," works great. So-called "bell wire" also works. String the wire up as high above the ground as you can get it. Connect the far end of the wire to a good ground rod driven into the soil to keep electrostatic charges from building up on the wire. (If you use a copper ground rod, the best kind, make sure it isn't too close to the roots of any trees or shrubs. Copper can kill plants.)

You can bring the near end of the wire in through a window sash and connect it to the back of your radio (after stripping off the insulation for an inch or so), where you should find a terminal specifically meant for external antennas. If your radio lacks an external antenna terminal and uses an attached collapsible whip antenna instead, strip 3 or 4 inches of insulation off the end of the antenna wire, collapse the whip all the way down except for an inch or two, and wrap the bare conductor around the end of the whip.

You can get enhanced performance if you use a small antenna tuner (also called a *transmatch*) intended for low-power HF ham radio use. A good source for these devices is a company called MFJ Enterprises. I recommend the model MFJ-16010 or a similar model. It'll ensure that you get the best possible *impedance match* between your random wire and your radio. Visit MFJ on the Web at

www.mfjenterprises.com

Never forget to disconnect your outdoor antenna from your radio when you aren't using the radio. Remove the end of the antenna entirely from the house, and lay it on the ground where nobody can trip over it or otherwise get in trouble with it.

Sunspots and Shortwave Radio

The sun's surface hosts violent storm-like phenomena that we see as *sunspots* because they radiate less visible light than the rest of the solar disk. You can view sunspots through a telescope equipped with a projection system or a dark filter over the objective lens or opening (never in the eyepiece). A typical sunspot measures farther across than the diameter of the earth! The biggest ones can grow to several earth diameters from one edge to the other. The darkest part of the spot lies at, or near, the center, a region called the *umbra*. The brighter zone around the umbra constitutes the *penumbra,* as shown in Fig. 2-5.

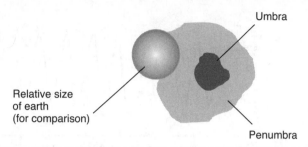

Umbra

Relative size
of earth
(for comparison)

Penumbra

FIGURE 2-5 A typical sunspot has a diameter greater than that of the earth.

Warning! Never view the sun directly through dark film negatives, thinking that the unexposed film will protect your eyes. Ultraviolet rays can penetrate the "celluloid" and do permanent damage to your vision. If you must use a "sun filter," use one that blocks ultraviolet rays. Don't use filters that fit in telescope eyepieces; the focused sunlight can heat the filter glass to the point where it breaks, giving your retina a horrific jolt. (As a child, I experienced that mishap and practically fell over backwards! Luckily, I did not suffer permanent eye damage.) Ideally, you should avoid direct sun viewing in any form, and use a telescope projection system that casts an image of the sun on a flat screen.

The Sunspot Cycle

For centuries, astronomers have known that the number of sunspots changes from year to year, month to month, and day to day. Short-term variations can be sporadic, but the long-term fluctuation has a regular and dramatic up-and-down rhythm known as the *sunspot cycle*. It has a period of approximately 11 years from peak to peak. The actual maxima and minima vary from cycle to cycle. Figure 2-6 shows the sunspot cycles since the year 1920 as a *very* approximate graph of relative activity versus time. At the time of this writing (2014), the sunspot cycle had passed a peak that was less intense than the several peaks that preceded it.

The amount of radio noise emitted by the sun is called the *solar radio-noise flux*, or simply the *solar flux*. Astronomers and communications engineers monitor the solar flux at a wavelength of 10.7 centimeters, corresponding to a frequency of 2800 MHz. At this frequency, the troposphere and ionosphere have minimal effect on radio waves, making observation of extraterrestrial RF phenomena easy. The 2800-MHz solar flux correlates with the sunspot cycle. On the average, the solar flux rises highest near the peak of the sunspot cycle, and dips lowest near sunspot minima.

Maximum Usable Frequency

Imagine some arbitrary medium or high frequency at which communication is possible between two specific points. Now suppose that you increase the frequency

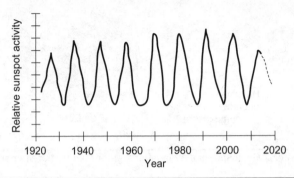

FIGURE 2-6   Approximate graph showing the sunspot cycles that have taken place since 1920.

gradually until communication fails. This cutoff point is called the *maximum usable frequency*, or MUF, for the two locations in question.

The MUF depends on the locations and separation distance of the transmitter and receiver. The MUF also varies with the time of day, the season of the year, and the sunspot activity. All of these factors affect the ionosphere. For a given signal path, the propagation usually improves as the frequency rises toward the MUF; at frequencies above the MUF, the "connection" deteriorates.

When you operate a communications circuit slightly below the MUF and then the MUF suddenly drops, a rapid and complete signal fadeout occurs. Long-time users of the HF radio spectrum know all about this effect. If you've been a ham radio operator for more than a couple of years, you've doubtless witnessed events of this sort. Hams will say that "the band is going out," or they'll talk about "QSB" (fading).

Tech Tidbit

When few or no sunspots exist, the MUF can drop down to about 4 or 5 MHz because the ionization of the upper atmosphere is least dense when sunspot activity reaches its minimum. At or near a sunspot cycle maximum, the MUF rises highest because the earth's upper atmosphere attains peak levels of ionization as a consequence of the high solar flux. In a few cases, ham radio operators and serious radio hobbyists have seen the MUF get up to 70 MHz or so! The number of sunspots, averaged over periods of a few days, correlates with the solar flux, which directly affects the degree to which the upper atmosphere gets ionized. That's why radio hams look forward eagerly to periods of high solar activity.

Solar Flares

A *solar flare* is a particularly dramatic eruption on the sun. When viewed through a telescope equipped with the appropriate filters, a solar flare appears as a bright fleck or streamer on the solar disk, thousands of kilometers across and thousands of kilometers high. At any given radio frequency, the solar flux increases abruptly when a solar flare occurs. This effect makes the solar flux useful for propagation forecasting: A sudden increase in the solar flux indicates that ionospheric propagation conditions will likely deteriorate within a few hours. Solar flares can occur at any time, but they take place most often near sunspot cycle peaks.

Solar flares emit countless high-speed atomic particles that hurtle through space at a small fraction of the speed of light, usually reaching the earth several hours after the occurrence of the flare. Because the particles carry an electric charge, they converge toward the earth's magnetic poles, causing a disturbance in the earth's magnetic field known as a *geomagnetic storm*. When that happens, you can see the aurora (northern and southern lights) on clear nights at high latitudes, and shortwave radio reception can fail completely within seconds at all frequencies. Even wire communications circuits sometimes suffer ill effects. In the extreme, the electric utility grid and the Internet can malfunction as well. Scientists don't know the exact mechanism that causes solar flares, but they can use these events to predict geomagnetic storms.

Ponder This!

In recent years, as our utility and communications infrastructures have grown more sophisticated, they've also become more "electromagnetically fragile" and vulnerable to the effects of geomagnetic storms. The news media have picked up with a vengeance on this trend and its possible consequences. Will society someday suffer a massive utility failure as a result of a solar flare? I don't know. I guess we can only wait and see, just as we did with, say, the Y2K (year-2000) computer bug! I don't sense much interest among government officials to spend any money to harden our infrastructure against this threat, however great or small it ultimately proves to be when the sun finally decides to have "the fit of all fits."

Longwave Radio

Communications specialists call the range of radio frequencies extending from 30 kHz to 300 kHz the *low-frequency* (LF) or *longwave band*. Informally, you can think of it as extending from the lowest allocated frequency in the United States at 9 kHz up to the low-frequency end of the standard AM broadcast band at 535 kHz. The wavelengths decrease from 10 kilometers down to 1 kilometer as you go up in frequency from 30 kHz to 300 kHz range. For this reason, some people call LF radio signals *kilometric waves*.

Waveguide Propagation

The ionosphere's E and F layers return all LF signals to the earth, even those waves that emanate straight up from the transmitting antenna. Likewise, LF signals of extraterrestrial origin can't reach us at the earth's surface because the ionosphere reflects them back into space. In effect, the ionosphere acts as a "radio wave mirror" at these wavelengths.

If you go down low enough in frequency, say to 100 kHz or below, the wavelength gets so long that the earth and the ionosphere combine to act like the surfaces of a transmission line known as a *waveguide*. Communications engineers and radio hams use waveguides at ultra high and microwave frequencies (300 MHz and above) to carry signals from transmitters to antennas, and/or from antennas to receivers. Physically, a transmission-line waveguide comprises a hollow metal duct or pipe, usually with a rectangular cross section. These transmission lines work with signals whose wavelengths measure millimeters or centimeters. Ionospheric waveguide propagation works with signals whose wavelengths measure kilometers.

In the waveguide mode of propagation, the transmitting antenna, which must produce a vertically polarized EM wave, delivers energy into the space between the earth and ionized layers of the upper atmosphere. The receiving antenna picks up the energy in a manner similar to an RF probe in an electronics lab. The *earth-ionosphere waveguide* has a minimum-frequency cutoff (lower limit) of around 9 kHz. At lower frequencies, signals don't propagate well because the wavelength is too long to let the earth and the ionosphere function properly as a waveguide. In fact, the earth-ionosphere waveguide effectively shorts out EM fields at frequencies below about 9 kHz, and that's why those frequencies remain largely ignored for use on this planet. (At the time of this writing, frequencies below 9 kHz were not legally allocated for any purpose in the United States.)

Waveguide propagation occurs along with surface-wave propagation, which you learned about in Chap. 1.

Tech Tidbit

Space travelers of the future might take advantage of surface-wave propagation at long wavelengths on planets that have no ionosphere, even though waveguide propagation won't work on such planets. Even on the moon, which has no ionosphere to return radio waves to the surface or to act with the surface to form a waveguide, LF waves might propagate for over-the-horizon communication by means of the surface wave. If a planet lacks an ionosphere, communication might work down to extremely low frequencies, indeed—even below 1 kHz—if the planet conducts electric currents well enough because the waveguide effect won't short out EM fields no matter how long the wavelength gets.

Maximum Usable Low Frequency

For any two points separated by more than a few kilometers, radio communication works well only at certain frequencies. At the longest wavelengths in the LF band, and also in the very-low-frequency (VLF) band below 30 kHz, communications circuits will almost always function worldwide, as long as the transmitter puts out enough RF power and the receiver has optimum selectivity and noise rejection capability.

Imagine a working communications circuit at a frequency of a few kilohertz between two specific points on the earth's surface. Suppose that you gradually raise the frequency. If the transmitter and the receiver are more than a few kilometers apart, the *path loss* (the difficulty with which signals propagate by means of surface-waves or waveguide mode) increases as the frequency goes up. This effect occurs because the lowest layer of the ionosphere becomes less reflective and more absorptive as the frequency rises, and also because the ground becomes less conductive so that the surface wave can't propagate as well. Under some conditions, you'll encounter a frequency at which communication deteriorates to the point of uselessness. This frequency will lie far below the useful shortwave communications range between the same two points. Engineers call this point in the LF band the *maximum usable low frequency* (MULF).

The MULF differs from the shortwave MUF. If you raise the frequency above the MULF, communication will not take place for a while, but if you keep going, you'll establish a working circuit once again (between the same two points) at a frequency called the *lowest usable high frequency* (LUHF). As the frequency increases still more, as you've learned, the circuit will fail again at the shortwave MUF.

How to Get Down There

If you want to listen to signals at frequencies below the standard AM broadcast band (535 kHz in the United States), you'll need a receiver that can tune through that range. Most older shortwave radios can't do it. For example, my 1960s Hallicrafters radio would only go down to the bottom of the AM broadcast band. However, many modern radios can "hear" almost all the way down to the audio frequency (AF) zone, which tops out around 20 kHz. A typical lower limit for such receivers is 30 kHz, the bottom of the LF band, corresponding to a free-space wavelength of 10 kilometers.

My own ham radio transceiver (Fig. 2-7) can receive signals down to 30 kHz and also up to 174 MHz, a frequency that lies well into the VHF part of the radio spectrum. (It can transmit only within the Amateur Radio bands that existed in the United States at the time of its manufacture around 2001.) Most other good HF ham "rigs" can do the same thing. It's a great feature, because you don't have to buy a separate shortwave radio. The versatility and signal-refining capabilities of a good HF ham radio compare favorably with the best dedicated allwave receivers.

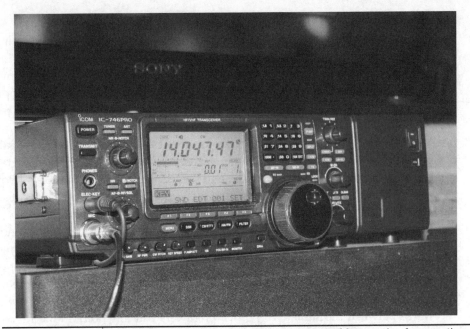

FIGURE 2-7   Most Amateur Radio HF transceivers cover a wide range of frequencies for reception with precision and flair! Here's an example: My own fixed-station ham radio set, which can tune continuously from 30 kHz to 174 MHz.

Whatever type of radio you use, you'll need a good antenna in order to effectively receive signals at long wavelengths. In general, as the wavelength gets longer, efficient transmitting antennas grow larger in commercial applications. There's no way around that conundrum: a decent radiating element for LF use must be large, indeed, and high up off the ground to boot. For receiving, things are a lot simpler. You'll find two types of receiving antennas in common use at LF and also at VLF:

1. A *random wire*, of indeterminate length (but as long as possible), strung up outdoors as high above the ground as you can place it; or
2. A *small loop* antenna with a tuning capacitor or preamplifier to boost the signal level, used indoors next to the radio.

The first type of LF and VLF listening antenna, the outdoor long wire, pretty much explains itself in terms of design. If you have a lot of real estate available (for example, if you live on a farm or ranch), consider yourself blessed! Put the antenna up between the tallest supports you can find, such as big trees. If you don't have any big trees on your property, you can put up masts or towers to serve as antenna supports. You can string the wire up between more than two supports, and it doesn't have to go in a straight line. The most important consideration is that your wire sit up in the clear, away from obstructions, and that you avoid placing it in such a way that it can fall or blow down on an electric utility line, or that an electric utility line can fall or blow down on it. You can run the wire down into your house through an open window, slam the window shut on the wire, and then connect the end of the wire to the antenna terminal on the back of the radio.

Warning!   Any outdoor antenna presents a danger when thunderstorms occur in your vicinity. As you make your wire antenna longer, the danger increases because it will readily accumulate massive electrostatic charges. A storm doesn't have to be overhead in order to put your radio, your home, and your life at risk. Even when a thunderstorm is several kilometers away, it can "charge up the atmosphere" enough to electrify a long wire antenna to an astonishing extent. You must connect your antenna to a good ground, preferably a long copper rod (2 meters or more) driven into good soil and placed at least 10 or 15 meters from your house, whenever you aren't using the radio. When you do use it, make sure that no thunderstorms exist anywhere nearby. You might want to keep a computer on one of the Internet radar sites so that if a storm approaches or rapidly develops, you can "see" it coming.

The second type of LF and VLF listening antenna in common use, the small loop, works according to the principle of "grabbing the magnetic component" of the EM field. Figure 2-8 shows the basic configuration. It should have an overall circumference of less than 0.1 wavelength (that's right, less!) at the highest operating frequency. It can be circular or square in shape. Such an antenna exhibits a sharp *null* (zone of minimum sensitivity) along the loop axis. At LF and VLF, a small loop must contain numerous turns of wire; as the frequency goes down, the number of required turns goes up. A variable capacitor, in conjunction with the coil, forms an *RF tuned circuit*. You set the loop's optimum receiving frequency by adjusting the capacitor.

For compactness and convenience, and also to increase the loop inductance, a length of insulated or enameled wire can be wound around a rod-shaped *inductor core* made of powdered iron. This type of antenna, called a *loopstick*, has directional characteristics similar to the ordinary small loop. The sensitivity is maximum off the sides of the coil, and a sharp null occurs off the ends. Instead of (or in addition to) using a tuning capacitor, you can insert a tunable RF preamplifier between the loop and the receiver's front end or antenna input. That device will give you some extra sensitivity and selectivity for bringing in the weakest signals and reducing interference from the strongest ones.

LowFERs

If you have a lot of patience, a sensitive LF receiver, and a well-engineered receiving antenna in a good location, you might hear signals from people who engage in a hobby called *LowFER* (an acronym that stands for *low frequency experimental radio*)

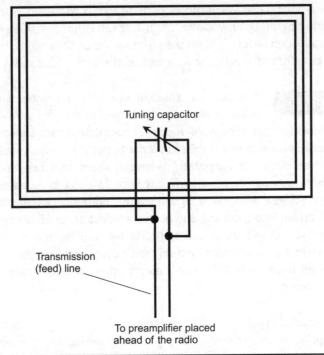

Tuning capacitor

Transmission
(feed) line

To preamplifier placed
ahead of the radio

FIGURE 2-8   A small loop antenna for receiving LF signals. The loop contains several turns and
can have a diameter of about 1 meter. The variable capacitor allows you to tune the
system to resonance (optimum signal conditions) at the desired frequency.

in the longwave frequency range between 160 kHz and 190 kHz. The actual
wavelengths go from 1875 meters down to 1579 meters, and you'll hear enthusiasts
refer to this slot of the spectrum in general terms as the *1750-meter band*. In addition
to checking out the experimenters on 1750 meters, you might tune down in frequency
a little ways and hear ham radio experimenters testing low-power transmitters on
136 kHz (2206 meters).

You don't need a license in the United States in order to conduct LowFER
activities that involve transmitting between 160 kHz and 190 kHz, as long as you
adhere to certain restrictions. (You must have a ham radio license to conduct
experiments that involve transmitting on 136 kHz.) For LowFER operation, you
can't use a transmitting antenna with an overall length that exceeds 15.24 meters,
including the transmission line between the radio and the radiating element itself.
In addition, your transmitter can't have an RF power output that exceeds 1 watt.
On account of these severe restrictions, most LowFERs use Morse code or digital
modes for communication in order to ensure that people listening for them get to
take advantage of the best possible signal-to-noise (S/N) ratio.

If you get lucky, you might hear a LowFER or two sending experimental signals
from a few hundred kilometers away. Maybe you'll want to get on the air as a

LowFER and see if you can make two-way contacts with other LowFERs! For more information about this hobby, visit the website of the *Longwave Club of America* (LWCA) at

www.lwca.org

Tip

If you want to get the most out of your LF and VLF listening experience, you'll need a receiver with a product detector, so that it can resolve digital and Morse code signals. If your receiver has only an envelope detector for AM, you won't hear much below the standard AM broadcast band because AM takes up too much bandwidth to make practical sense at LF and VLF.

Converters for VLF

If you want to listen to signals at frequencies below 30 kHz and your receiver can't do it, you can build a converter that will let you listen all the way down to DC (0 Hz), at least in theory. (In practice, noise and the converter's own oscillator make reception difficult below about 3 kHz.) Figure 2-9 is a generic diagram of a *VLF converter* that mixes incoming VLF signals with an unmodulated carrier from a local oscillator at some frequency in the shortwave band, such as 3500 kHz. The circuit produces, in effect, a wideband AM signal with sidebands that extend several tens of kilohertz above and below that frequency. You can use a shortwave radio receiver to tune through those sidebands and hear VLF signals at the sum and difference frequencies. For example, a VLF signal coming in at 17 kHz would show up in the receiver at 3517 kHz (3500 + 17) and also at 3483 kHz (3500 − 17).

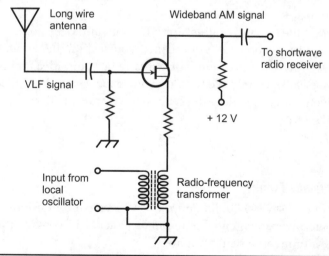

FIGURE 2-9   Generic diagram of a VLF converter. Exact component values must be determined by experimentation.

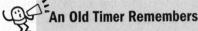

An Old Timer Remembers

Way back in the 1970s when I attended college in Minneapolis, Minnesota and visited my parents' house down in Rochester every other weekend or so, I had a great transmitter and receiver for ham radio. The transmitter was a Drake T-4X and the receiver was a Drake R-4A. They worked entirely (or almost entirely) with vacuum tubes. The transmitter had a "spot" setting that activated all the stages except the RF driver and final amplifier, so I could find and set my exact transmitting frequency by listening to my own signal in the receiver.

One day I decided to see if I could use my ham rig as a VLF converter by plugging my antenna (a vertical on the roof, only 17 feet high, meant for the 14-MHz ham radio band) into the microphone jack of the transmitter, and then transmitting an AM signal in the "spot" mode. I imagined that I could then set the transmitter to some known frequency, say 3500 kHz, and tune upward from there to hear the sidebands that might get generated by VLF signals at the microphone input. I figured it constituted a "long shot," but adverse odds never daunted me prior to that time (nor have they since).

When I finally carried off the experiment, which required only the purchase of an extra microphone plug, my reward came instantly. I heard signals all through the VLF band, with the lowest ones around 13 kHz. The trick worked all the way up to about 60 kHz, beyond which point the audio circuits in the T-4X didn't work very well. Below about 6 or 8 kHz, noise from the utility lines made reception impossible, but I did not expect any signals to exist there anyway, since the FCC had made no formal allocations below 9 kHz. The trick did not work in the SSB mode because the filters in the RF stages of the T-4X cut off above 3 kHz. No such filters existed, evidently, in the AM mode.

Here's the "moral" to this story: If you want to try something, no matter how high the theoretical odds against you seem to be, go ahead. You never know what will happen until you actually test your idea and find out. By the way, this isn't the only weird and exotic experiment that I did back in those (relatively) carefree college days. But you've probably figured that out already, haven't you?

Not a Tweak Freak?

You can find several "build and play" VLF converter designs and even some commercially manufactured units online. Enter "VLF converter" into the phrase box of your favorite search engine.

Dirty Electricity

As you learned in Chap. 1, electric and magnetic flux lines surround all current-carrying wires. Around a straight wire, the electric or E flux lines run parallel to the wire, and the magnetic or M flux lines encircle the wire, as shown in Fig. 2-10A. If the wire carries constant DC, the E and M fields remain constant and steady. If the wire carries AC or pulsating DC, the fields fluctuate, producing an EM field that emanates from the wire. At great distances from the wire, the E and M lines of flux are perpendicular to each other in a nearly flat plane, forming an EM field that travels through space at the speed of light, as shown in Fig. 2-10B. This field induces AC in any nearby object that conducts electricity, such as your radio antenna. Unfortunately, any nearby AC utility line or house wiring produces EM energy not

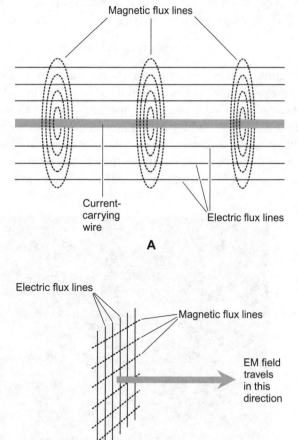

FIGURE 2-10   At **A**, the electric (E) and magnetic (M) lines of flux around a straight, current-carrying wire. At **B**, the flux lines far away from a current-carrying wire.

only at 60 Hz, but also at harmonics (whole-number multiples) of 60 Hz, as well as at other unrelated frequencies. The spectrum below a few kilohertz suffers from "EM pollution" on account of this *dirty electricity* in any area served by an electric utility. At LF and especially at VLF, you'll notice the effects as soon as you tune a radio down there. The EM waves that you want to hear, and the ones that you don't want to hear, both originate and propagate in the same way, as Figs. 10A and 10B show. In a city, dirty electricity can render VLF reception impossible. Not only can you hear this stuff, but also, with the appropriate equipment, you can see it too! Here's how you can put together a device that will detect dirty electricity and produce a graphic display of its characteristics (although it won't help you get rid of it).

A Cool Little Program

A simple freeware program called *DigiPan*, available on the Internet, can provide a real-time, moving graphical display of EM field components at frequencies ranging from 0 Hz (that is, DC) up to around 5500 Hz. Here's the website:

www.digipan.net

DigiPan shows the frequency along a horizontal axis, while time is portrayed as downward movement all across the whole display. Figure 2-11 shows this scheme. The relative intensity at each frequency appears as a color. You can adjust the colors to suit your taste. If no energy exists at any frequency in the program's range, the display will appear completely black. If there's a little bit of energy at a particular frequency, you'll see a thin, vertical blue line creeping straight downward (if you leave the program set for its default color scheme). If there's a moderate amount of

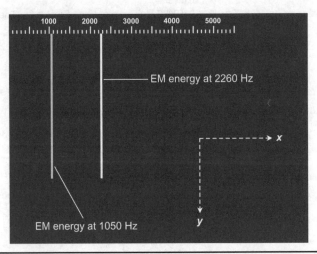

Figure 2-11   Basic geometrical scheme for the DigiPan display. The horizontal axis portrays frequency. Time "flows" downward across the whole screen. Signals show up as vertical lines. This drawing shows two hypothetical examples.

energy, the line turns yellow. If there's a lot of energy, the line becomes orange or red. The developers and users of DigiPan and similar programs have coined the term *waterfall* for the display, because of its appearance when signals exist at numerous frequencies.

Did You Know?

The originators of DigiPan wrote it for digital communication in *phase-shift keying* (PSK), a popular mode among ham radio operators. However, DigiPan can also function as a *baseband spectrum monitor*, showing the presence of AC-induced EM fields, not only at 60 Hz (which you should expect) but also at many other frequencies (which you might not expect until you see the evidence). If you have a good Internet connection, DigiPan will download and install in a minute or two on any computer that runs Windows XP, Windows 7, or Windows 8, although some of the program's more exotic features won't run on Windows 7 or 8.

For Nerds Only

DigiPan isn't the only program you can use to display dirty electricity and other EM spectral effects. Another program, called *HamScope*, works well too. Go into your favorite Internet search engine and enter the single word "HamScope" to find download sites. This program, like DigiPan, has a small "footprint" on a personal computer, and will download fast. As for learning to use it, you can do as I did: Follow the COEISASWH paradigm (click on everything in sight and see what happens).

The Hardware

To observe dirty electricity on your computer, you'll need an "antenna" that can "pick up the trash." Get an audio cord at least 3 meters long with a mini two-wire (monaural) phone plug on one end and spade lugs on the other. Cut off the U-shaped spade lugs from the audio cord with a scissors or diagonal cutter. Separate the wires by pulling them apart along the entire length of the cord so that you get a phone plug with two wires attached, forming a little dipole antenna.

Insert the phone plug into the *microphone* input of your computer. Arrange the two wires so that they run in different directions from the phone plug. You can let the wires lie anywhere, as long as you don't trip over them! This arrangement will make the audio cord pick up EM energy.

Open the audio control program on your computer. If you see a microphone input volume or sensitivity control, set it to maximum. Set your computer to work with an external microphone, not the internal one. If your audio program has a "noise reduction" feature, turn it off. Set the microphone gain (or input sensitivity) to maximum. Then launch DigiPan and follow these steps, in order.

- Click on "Options" in the menu bar and uncheck everything except "Rx."
- Click on "Mode" in the menu bar and select "BPSK31."
- Click on "View" in the menu bar and uncheck everything.

Once you've carried out these steps, the upper part of your computer display should show a jumble of text characters on a white background. The lower part of the screen should be black with a graduated scale at the top, showing the numerals 1000, 2000, 3000, and so on. Using your mouse, place the pointer on the upper border of the black region and drag that border upward until the white region with the distracting text vanishes.

If things work correctly, you should have a real-time panoramic display of EM energy from zero to several thousand hertz. Unless you're in a remote location far away from the utility grid, you should see vertical lines of various colors. These lines represent EM energy components at specific frequencies. You can read the frequencies from the graduated scale at the top of the screen. Do you notice a pattern?

Harmonics

A pure AC sine wave appears as a single *pip* or vertical line on the display of a spectrum monitor (Fig. 2-12A). This pip means that all of the energy in the wave is concentrated at one frequency, known as the *fundamental frequency*. But many, if not most, AC utility waves contain *harmonic* energy along with the energy at the fundamental frequency.

A harmonic is a whole-number multiple of the fundamental frequency. For example, if 60 Hz is the fundamental frequency, then harmonics can exist at 120 Hz, 180 Hz, 240 Hz, and so on. The 120-Hz wave is at the *second harmonic*, the 180-Hz wave is at the *third harmonic*, the 240-Hz wave is at the *fourth harmonic*, and so on. In general, if a wave has a frequency equal to n times the fundamental frequency where n is some whole number, then that wave is called the *nth harmonic*. (The fundamental frequency is the first harmonic by definition). In Fig. 2-12B, a wave is shown along with its second, third, and fourth harmonics, as the entire "signal" would appear on a spectrum monitor.

When you look at the EM spectrum display from zero to several thousand hertz using DigiPan, you'll see that utility AC energy contains not only the 60-Hz fundamental, but also *many* harmonics. When I saw how much energy exists at the harmonic frequencies in and around my house, my astonishment knew no bounds. I had suspected "dirt in the ether," but not *that* much! Figure 2-13 shows my own DigiPan display of dirty electricity from 60 Hz to more than 4000 Hz in the workspace where I wrote this book. Each vertical trace represents an EM signal at a specific frequency. If the electricity were "perfectly clean," you'd see only one bright vertical trace at the extreme left end of the display.

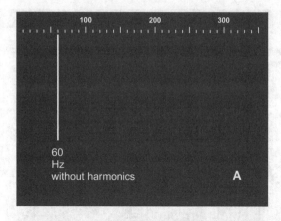

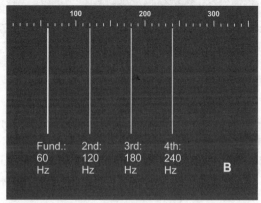

Figure 2-12 At **A**, a spectral diagram of pure, 60-Hz EM energy. At **B**, a spectral diagram of 60-Hz energy with significant components at the second, third, and fourth harmonic frequencies.

Try This!

Place a plug-in type vacuum cleaner near your EM pickup antenna. Switch the appliance on while watching the DigiPan waterfall. When the motor first starts up, do curves suddenly appear on the display, veering to the right and then straightening out as vertical lines? Those contours indicate energy components that increase in frequency as the motor "revs up" to its operating speed and maintain constant frequencies thereafter. When the motor loses power, do the motor's vertical lines curve back toward the left before they vanish? Those curves indicate falling frequencies as the motor slows down. Try the same tests with a hair dryer, an electric can opener, or any other appliance that plugs into a wall outlet and contains an electric motor. Which types of appliances are the "cleanest"? Which are "filthiest"?

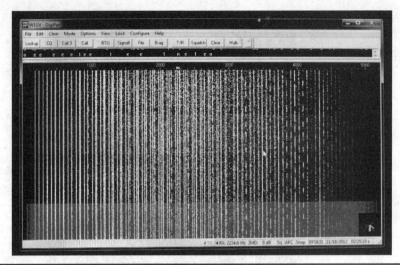

FIGURE 2-13   Dirty electricity in my work space, as viewed on DigiPan.

Fact or Myth?

If you conduct in-depth, Internet-based research into this subject, you'll come across a lot of sites that warn about health hazards posed by dirty electricity. Some sites will give you case histories, horror stories, and wild tales about illnesses, pains, and cancer, and attribute all of the trouble to dirty electricity. How serious is the danger, really? As a former radio-frequency (RF) engineer and antenna specialist, the best answer I can give you is "I don't know." I suspect, however, that if the "dirt" in dirty electricity has adverse health effects, the "clean" part, which produces far stronger EM fields, probably does as well. So any efforts to "clean up" electricity, in the hopes of getting rid of its potential ill effects, are futile, unless we as a society decide that we don't want to live with utility-provided electricity anymore!

Try This Circuit for Relief!

You might be able to reduce the noise from dirty electricity at frequencies below about 100 kHz by using two antennas that receive the noise at equal amplitudes but receive the desired signals at different amplitudes. You must connect the two antennas together so the noise signals cancel each other out. Figure 2-14 shows an example of a circuit that I have used to accomplish this feat. With a little testing and tweaking, maybe you can make it work too.

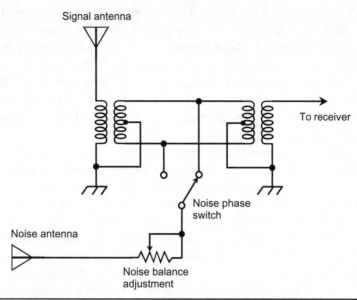

FIGURE 2-14   A noise-canceling antenna scheme that works especially well at VLF.

In order for a circuit of this sort to function at its best, the signal antenna and the noise antenna must lie reasonably near each other, so they pick up local humanmade noise whose characteristics are as nearly identical as possible. But they should not be too close together, or the desired signals will get canceled along with the noise. Place the signal antenna in a location favorable for the reception of the stuff you want to hear, and locate the noise antenna so that it will receive as much dirty electricity as possible (and relatively little of the desired signals). For example, the signal antenna might be a tall vertical element operating against an earth ground; the noise antenna can comprise a short, random length of wire run on the floor inside your house.

The potentiometer lets you adjust the level of signals and noise from the noise antenna, while having a minimal effect on the overall signal level from the two antennas combined. The phase switch allows for injection of the noise from the noise antenna either "right side up" or "upside down," so that you can make sure that it enters the rest of the circuit in *phase opposition* with the noise from the signal antenna. At frequencies below approximately 30 kHz, where wavelengths are measured in kilometers, the noise from the two antennas will be either coincident in phase or opposite in phase. After you have found the correct switch position by trial and error, adjust the potentiometer until you get minimum "trash" in your receiver.

When I first put this circuit together, I didn't know how well it would work. I was astonished when I saw the noise go from over S-9 on my receiver signal meter down to only about S-3! That's a reduction of six S units, where each S unit (on most receivers) represents about 5 dB. If you take it as 5 dB, then you can see that I got a 30-dB improvement with the noise-canceling antenna system, which means that the noise power at my receiver's "antenna terminals" (T-4X microphone jack) went down to only 1/1000 of its previous level. The desired signals, however, remained as strong as ever.

I have used this scheme for receiving signals at VLF, in conjunction with a VLF converter and a shortwave receiver, with excellent results. The signal antenna was a 16-foot, ground-mounted vertical antenna fed with coaxial cable, intended mainly for operation on the 14-MHz ham radio band. The noise antenna was wire a few feet long, running under a rug on the hallway floor.

As you increase the frequency above the VLF band, this scheme gets less effective because the noise impulses at the two antennas are no longer exactly in phase coincidence or phase opposition unless the antennas are so close together that the circuit cancels the desired signals as well as the noise. In addition, I have found that this circuit does not work against sferics (thunderstorm "static").

The Whole Story

At the time of this writing, I found a website that provides tables that you can download, showing all of the frequency allocations in the entire radio spectrum. You can use this table as a guide as you navigate your way through the labyrinth of frequencies and wavelengths that a good allwave radio receiver can "hear," and then some! Go to

www.fcc.gov/encyclopedia/radio-spectrum-allocation

You can choose either Word or PDF formats for offline viewing. I downloaded the PDF version and printed it for permanent reference (or at least until the next change in official spectrum allocations). By the way, if you plan to print this document, get your printer ready to do some hard work. The PDF version tallies up with a 168-page count from the "top" to the "bottom"!

Which Radio's for You?

If you want to get into SWL or AWL, you can buy a ready-made receiver, buy a kit that you can assemble into a receiver, or build your own radio from scratch. The best commercially manufactured radios can run you upwards of a thousand dollars; the cheapest kits or "mini radios" might cost you only $30 or $40.

I won't attempt to make any product comparisons or recommendations here. Shortwave and allwave receiver designs and features keep changing, almost from day to day. By the time you read anything specific that I might quote, you'd have good reason to suspect that the information had already grown obsolete!

If you want to get into SWL or AWL with a reasonably good radio right away, I recommend that you go to your local Radio Shack store and shop around. If they don't have physical units right on their shelves, go to the Radio Shack website and you'll find a wide range of models, one of which will surely suit your desires! Their home page is at

www.radioshack.com

May I also suggest that you attend a meeting of a local ham radio club, or stop by one of the hundreds of ham radio conventions that take place nationwide? They'll provide all sorts of advice and insights— and they'll probably try to get you interested in their hobby as well. If you get interested in ham radio and decide to go for a license so that you can create signals as well as merely listen to them, trust me: You'll be glad you did.

CHAPTER 3

Ham Radio Communications Modes

Non-hams might think that Amateur Radio operators use only "code" or "voice," forgetting about other methods by which hams can, and often do, exchange information. Let's take a close look at the most common forms that ham radio communication takes.

Morse Code

The Morse code has existed longer than any other method of sending and receiving wireless messages. It came about in the early 1800s when engineers perfected the wire telegraph and could send only simple current pulses along the lines. In theory, Morse code, which hams call CW (for *continuous waves*), is a *binary digital* mode in which the signal always has either of two well-defined states: full-on (*key down* or *mark*) or full-off (*key up* or *space*). Hams send and receive CW as a sequence of short audio tones called *dots* and longer audio tones called *dashes*. So the waves aren't truly continuous; they're intermittent.

How Fast (or Slow) Does It Go?

Radio hams express code speed in words per minute (wpm), where one "word" averages five characters in length with a three-bit pause between characters. Some engineers define CW speed as the number of times that the word "Paris," along with the seven-bit pause afterwards, repeats in one minute when sent over and over in perfectly formed code. The word "Paris" plus the seven-bit pause has a total of 50 bits, so 1 wpm works out to 50 bits in 60 seconds, or *5/6 bits per second* (5/6 bps).

If you want to translate the speed of a transmission from words per minute into the equivalent speed in bits per second, multiply by 5/6. Once you do that calculation a few times, you'll realize how slowly ham radio CW conversations actually go, relatively speaking. For example, 12 wpm = 10 bps, and 30 wpm = 25 bps. For Morse code, 30 wpm represents a respectable speed. Even the best operators, who can send and read CW at around 90 wpm, communicate at only 75 bps. Compare that rate with the computer *modem* (modulator/demodulator) speeds that we measure these days in millions of bits per second, or in some cases, billions of bits per second!

As slow as it pokes along compared with computer data transmissions, some ham radio operators, myself included, enjoy CW as an art form; it's just plain fun to send and receive. Well-sent code sounds rhythmic, almost mesmerizing, to a true CW buff. You don't have to know the code to get a ham radio license now, although at one time not only did you have to know the characters but also you had to demonstrate proficiency at specific speeds, both sending and receiving, depending on the class of license that you sought.

Reading and Copying It

If you want to learn the code and boost your skill once you know the characters, you'll find no substitute for practice—hours and hours of it! The official station of the American Radio Relay League (ARRL), bearing the call sign W1AW, sends code practice sessions regularly on some of the HF ham bands at speeds ranging from 5 wpm to 35 wpm. They've been doing it for decades. I took advantage of those sessions to get my speed up to the 13-wpm level that I had to demonstrate during my General Class license test in 1967, and also to get up to the 20-wpm level that I needed in order to pass the Extra Class exam in 1973.

For a schedule of the W1AW transmissions in all modes, including code practice and official bulletins of interest to radio hams, visit the ARRL website at

www.arrl.org

The best way to learn the code, and get better at it once you know it, involves writing down received characters on paper, an activity called *copying CW*. I still copy some code every now and then to see if I can put down a solid minute of characters, error-free, at 20 wpm after all these years. I can do it if I'm in a good mood, so I still feel "qualified." In fact, on a particularly fine day, I can manage 25 wpm! I recommend that you use a college-ruled spiral notebook along with a reliable medium-point roller pen that won't smear. Write your text on every other line. Once you've put yourself through this exercise for a few hours, you'll know if you're "cut out" for CW. Most people aren't; and in fact, they hate it. I, along with a (dwindling) few others, love it.

Code Requirement: Good or Bad?

As I mentioned a few moments ago, getting a ham radio license once required knowledge of, and proficiency in, the Morse code. When I first became a ham in 1966, I got what they called the *Novice Class license*, which required me to send and copy a solid minute of characters, error-free, at 5 wpm. Even the Technician Class license for experimenters at the higher frequencies carried a 5-wpm requirement. Gradually, the Federal Communications Commission (FCC) eased the copying standards for United States ham radio operators, and ultimately they abolished the code requirement for all classes of license. Some Amateur Radio old timers will tell you that this evolution of events has harmed ham radio, bringing in "low quality operators." I think that the opposite has happened. People who used to dismiss ham radio because of the code requirement are now getting licenses, and some of them are mighty smart future engineers!

Sending It

Most people who've had experience with the Morse code will tell you that learning to send it comes easier than learning to read or copy it. That was my experience! You can learn to send code on a *straight key*, also called a *telegraph key*, with which you make the dots and dashes by pushing down on a lever against a spring-tension bearing. Once you can send about 15 wpm that way, you can switch to an *electronic keyer*.

Electronic keyers come in two major varieties. One type uses a lever or pair of levers that you push towards the right to make dots, and toward the left to make dashes. Figure 3-1 shows one of the most popular of these devices, called a *paddle*. It has a single lever, the type that I prefer. It's manufactured by a company called Vibroplex, and in my opinion, they're the best! You can find their website at

www.vibroplex.com

The other type of keyer uses a keyboard. You simply type in what you want to send, and as long as you stay ahead of the output with your typing, the device will generate perfect code at whatever speed you select. These *keyboard keyers* first came out in the 1970s to meet the demands of high-speed CW operators who wanted to carry on conversations at 50 wpm or more. Figure 3-2 shows the keyboard keyer that I acquired around 1975, manufactured by a company called HAL Communications. That thing still works! However, most hams use computer programs nowadays to generate CW if they want to send it with a keyboard.

Figure 3-1   A paddle for sending Morse code with an electronic keyer. This model is the Vibroplex VibroKeyer Deluxe. It has existed for over half a century, and you can still buy one today.

Figure 3-2   A keyboard for sending high-speed code, an art that enjoyed popularity in the 1970s.

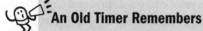
An Old Timer Remembers

During the early and mid 1970s when I attended college, I got interested in high-speed CW (also called *QRQ*, which literally translates to "Let's go faster"). A group of hams hung out on 40 meters around 7.030 MHz and had casual QRQ CW conversations at speeds in excess of 50 wpm. Sometimes they carried on their activity on 20 meters around 14.030 MHz. I won't identify any of them specifically, but if you want to know a little bit about these guys and gals, do an Internet search on the phrase "Five Star Operators Club." I never got up to Five Star status (80 wpm or more), but I recall having a *QSO* (conversation) with a California gal at 68 wpm. Yes, I used to measure my sending and receiving speeds to the nearest word per minute! I guess in retrospect that habit might have qualified as an obsession.

Radioteletype

Radioteletype (abbreviated RTTY and sometimes pronounced "ritty") involves the wireless exchange of printed or screen-displayed text matter. On the HF bands, hams use frequency-shift keying (FSK) to accomplish this. The mark and space frequencies differ by 170 Hz, resulting in the so-called *170-Hz shift*. Before about 1970, you could often hear *850-Hz shift* signals as well, but that mode has fallen from favor because it takes up more bandwidth than a ham-radio RTTY signal needs. On the VHF bands, audio tones are applied to the microphone input of an AM or FM transmitter, resulting in an emission mode known as *audio-frequency-shift keying* (AFSK).

Baudot Teleprinters

The oldest RTTY code is called *Baudot*, and allows for printing of uppercase letters of the alphabet, numerals, and common punctuation marks. A sequence of pulses called "line feed" causes the teleprinter carriage or computer display to back up and move to the next line before printing will resume.

Mechanical teleprinters, some of which remain in use today, employed either continuous-feed "roll paper" or continuous-feed "accordion fold" paper. When running at full speed, a Baudot teleprinter produced characters at a rate equivalent to 60 wpm. That's still the standard Baudot speed for ham radio use today. Another speed specification, called *baud*, expresses the number of signal-state changes (from mark to space or vice versa) per second. The typical ham radio RTTY speed, equivalent to 60 wpm in plain English text, is 45.45 baud.

Teleprinters were, in effect, automatic electric typewriters. In fact, when transmitting, you'd type on a keyboard just like the one on any common electric typewriter of the day. The mechanical system wouldn't let you pound out characters at a rate any faster than the maximum 60 wpm. Most hams, being poor typists, didn't find that constraint troublesome. If you could type faster than 60 wpm and tried to do it on one of those machines, it would physically restrain you; you could feel the force in the keys pushing back on your fingers!

You can find quite a few hams communicating with RTTY using 45.45 baud Baudot today. They hang out in certain parts of the HF bands, mostly in the upper parts of the digital-mode segments, just "above" (in frequency) the Morse code gang. The ARRL Headquarters station W1AW still transmits official bulletins regularly using 45.45-baud Baudot. Relatively few people copy the signals on mechanical teleprinters nowadays; computers do the job quietly and without mechanical parts, ink cartridges, and paper to break down or run out.

ASCII

If you do serious work on RTTY, you'll find Baudot rather primitive because you can't transmit and receive lowercase letters. In addition, the Baudot non-alphanumeric symbol set is limited. Engineers invented the *American Standard Code for Information Interchange* (ASCII, an acronym pronounced "askee") in the middle of the last century to address this issue. This code comprises a lot more code elements than Baudot has. In fact, ASCII contains 128 different character codes, while Baudot has only 32.

When using ASCII with FSK or AFSK for RTTY, you can work at speeds higher than 45.45 baud (60 wpm). Common speeds are 75 and 100 wpm. However, ASCII is not used very often for plain old FSK-style RTTY by radio hams. Instead, ASCII is reserved for other digital modes we'll discuss later in this chapter. Some of those modes have the capability to go many times the speed of conventional RTTY, but some actually go slower. In any case, the allowable bandwidth is limited on the HF bands by FCC law, and that constraint places a practical maximum on the communication speed.

Stay Tuned!

At the time of this writing (late 2013), ham radio operators have begun to discuss whether or not the traditional bandwidth limitations for digital transmissions on the HF ham bands ought to be increased to allow for higher communication speeds. You'll find information about this trend, as events unfold, at the ARRL website. Once again, it's

www.arrl.org

The ARRL will publish information about any change in the law concerning allowable bandwidth, when and if it occurs.

Station Equipment for RTTY

If you want to start working RTTY right now and you don't want to take the vintage "teleclunker" route, you'll need a computer, and probably also a separate interface unit. The computer, with the proper program and audio input, can read and display RTTY on your monitor right out of the box. All you need is an audio cord to connect your radio's audio output (from the headphone jack or from an auxiliary connector on the rear panels of some radios) to the "line in" jack on your computer, along with one of the several RTTY programs available today.

If you want to get your radio to transmit RTTY, an outboard interface unit will help you do it with versatility and ease, although some computer programs claim to let you transmit all sorts of digital signals without one. I use an interface box called the RigBlaster Pro, manufactured by a company called West Mountain Radio, between my computer and my radio. I've never had any trouble with it after over a decade of use along with my Icom IC-746 Pro transceiver. At the time of this writing, West Mountain Radio maintains a website at

www.westmountainradio.com

To get an idea of currently available computer programs that can read and generate AFSK tones (which will come out of your radio when you receive, and go into it when you transmit, regardless of the actual RF signal frequency), do an Internet search on the phrase "programs for RTTY" and explore the hits that you get. I use a basic program called MMTTY, which came on a CD-ROM with my digital mode interface unit, manufactured by West Mountain Radio around 2003. You can also find MMTTY on the Internet, but you'll have to look around a bit to find the latest version. Figure 3-3 shows the full MMTTY display with the program

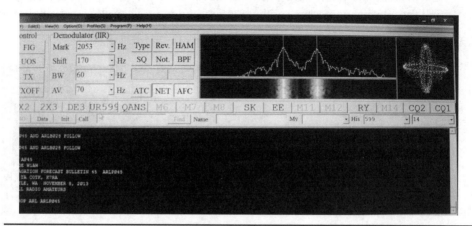

FIGURE 3-3   The MMTTY program display for Baudot includes a text display with adjustable background along with three different tuning scopes.

in action. This and most other RTTY programs include graphical tuning aids to help you properly adjust your radio's frequency.

Normally, if you want to receive RTTY, you should set your radio to lower sideband (LSB) mode. Some radios have a special RTTY filter that can give you a narrow passband to minimize interference from signals at nearby frequencies. Tune your receiver so that the mark signal produces an audio tone of 2125 Hz and the space signal produces an audio tone of 2295 Hz, representing a 170-Hz shift. (Those particular frequencies aren't an absolute requirement, but they've become the informal standard.) Some programs display the audio frequency numbers for you; with others, you'll have to guess, and that's where the graphical tuning aids really help.

Figure 3-4 shows a close-up of the three tuning aids in MMTTY. The display at the upper left indicates optimum tuning when the two signal peaks coincide with the vertical lines. The lower left display scrolls down with time; you should align the brightest parts of the signal with the two vertical lines above it. The right-hand display shows two ellipses that appear oriented horizontally and vertically, crossing each other as shown, when you have a signal tuned in correctly, and assuming that the signal has the proper amount of frequency shift. Any one of these three types of display will work satisfactorily for most operators.

Tip

Some digital communication programs are published as *freeware*, meaning that you can download and use them as much as you want, and for as long as you want, without paying anything. Others require a one-time payment, usually a modest amount, after a free trial.

I recommend that you test two or three RTTY programs on your computer to find the one that agrees best with your temperament. They all have a small resource

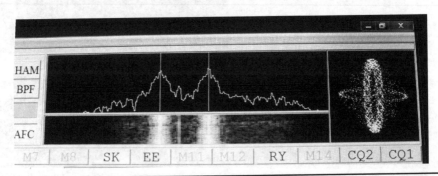

FIGURE 3-4 Closeup of the MMTTY tuning scopes. They're intuitive; playing with the program and your radio for a few minutes will teach you all you need to know about how they work.

footprint, meaning that they don't gobble up much memory space, hard drive space, or processing power on latter-day computers that typically have several gigabytes of memory, around a terabyte of available storage on the hard drive, and multiple-core processors. However, not all of these programs work to their full potential on newer operating systems, and a few of them might not work at all on some computers.

Most ham RTTY programs are written for Microsoft operating systems, notably the venerable old Windows 98 or Windows XP, so you'll face a challenge if you're a Mac user. However, I did locate a program called *MultiMode* after an Internet search, and the authors claim that it allows a Macintosh computer to send, as well as receive, RTTY and several other digital modes without a separate interface unit.

Patience Equals Power

At the risk of sounding a bit patronizing, I advise you to create a system restore point for your computer before you download any program, whether it's for ham radio or anything else. Download the executable file onto an external medium, such as a flash (thumb) drive, and then give it a virus scan with a good up-to-date antivirus program, before you install it or try to launch it. I have to remind myself of this precaution every time I download anything because sometimes I want to "cut to the chase" right away. Patience equals power!

Phase-Shift Keying

Several variants of RTTY emission exist besides FSK. One of the most popular of these modes, known as *phase-shift keying* (PSK), conveys text by varying the phase, rather than the frequency, of the carrier wave. Whenever you change the frequency of a signal, an inherent phase shift goes along with it, and vice versa. So in effect, PSK produces the same results as FSK.

Binary PSK

In ham radio, the most popular variant of PSK splits the carrier into two opposing phases, so it's sometimes called BPSK for *binary phase-shift keying*. The carrier phase remains constant during the entire length of any single bit, and might (or might not) shift by 180 degrees, or a half cycle, for the next bit. The most common BPSK mode works at 31.25 baud and goes by the full-length technical moniker BPSK31.

For hams who can type well and who adhere to common English usage in regards to capitalization and punctuation, plain text (such as what you're reading right now) BPSK31 signals come out at 45 to 50 wpm. Of course, for slow typists, the speed comes out slower than that. And, for those folks who insist on sending their text in all uppercase (as the old Baudot code did by default), the practical speed with PSK turns out closer to 30 wpm.

Where to Find It

Most BPSK31 enthusiasts populate informally agreed-on frequency slots in the digital-mode segments of the HF ham bands. These zones are 3 kHz wide, about the same as the bandwidth of an SSB signal. Look for BPSK31 activity in the following frequency ranges:

- 160 meters: 1.838 MHz to 1.841 MHz
- 80 meters: 3.580 MHz to 3.583 MHz
- 40 meters: 7.035 MHz to 7.038 MHz and 7.070 MHz to 7.073 MHz
- 30 meters: 10.140 MHz to 10.143 MHz
- 20 meters: 14.070 MHz to 14.073 MHz
- 17 meters: 18.100 MHz to 18.103 MHz
- 15 meters: 21.070 MHz to 21.073 MHz
- 12 meters: 24.920 MHz to 24.923 MHz
- 10 meters: 28.120 MHz to 28.123 MHz

Occasionally you'll hear a signal that's a long way outside these zones, and people often stray slightly above or below them, especially on 40 meters and 20 meters when activity grows heavy. In any case, as a general rule, BPSK31 signals always go in the digital-only sections of the ham bands.

Tip

At the time of this writing, I found an excellent website that delves into technical detail about PSK. It appeared at

http://www.arrl.org/psk31-spec

If you can't find it because the Web page location has changed (a common occurrence; oh, the wonders of the Web!), try a search on the word sequence:

PSK31 is a digital communications mode which is intended for live keyboard-to-keyboard conversations

---

Programs That Do PSK

Numerous radio hams have written programs for receiving and sending PSK in its various "flavors." All of these programs can decode and encode BPSK31. You can get some of them as downloadable freeware on the Internet. I like two in particular: DigiPan and HamScope. Both of them originated for Microsoft Windows XP, but I have run them on Windows 7 and Windows 8. My ham station computer has both of these programs installed. I use DigiPan's excellent high-definition tuning aid to see where CW signals reside within my receiver's passband, so I have the text screen and control bar disabled. For ordinary BPSK operation, I use HamScope because it can also decode signals in the exotic mode known as multiple-frequency-shift keying (MFSK).

DigiPan 2.0 was created by Amateur Radio operators. Figure 3-5 shows the operating screen. You can adjust the font style, font size, font color, and text background color, as well as the extent to which the text fills the screen. The lower part of the screen contains a tuning aid called a *waterfall*. All good PSK programs have waterfall displays, at least as an option. The name comes from the fact that the signals, which appear as bright vertical lines, move down the screen with the passage of time, so that the whole display looks like a real waterfall if you use your imagination. You tune the receiver within the passband by giving your mouse a single left click at the top of the bright colored line representing the signal you want to decode. When you want to transmit, you can hit a button marked "T/R" on a control bar at the top of the screen. (I have the control bar disabled in the rendition shown in Fig. 3-5. To display it, hit "View" and then hit "Control Bar.")

HamScope 1.56 was authored by another ham radio operator. Figure 3-6 shows the operating screen for HamScope as it receives some signals in the BPSK31 mode. You can select between a waterfall display and a *spectrum display* for tuning. I use the spectrum display in HamScope because it offers a clearer indication (to my eyes, anyhow) than the waterfall does in that particular program. The display shows signal amplitude versus frequency averaged over a short time interval as the upper

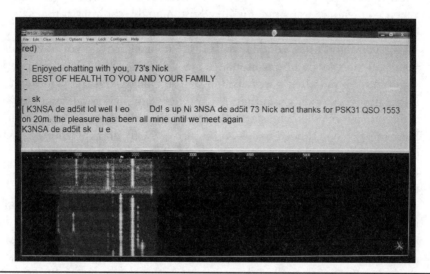

FIGURE 3-5   The DigiPan program working in BPSK31 with a waterfall tuning aid. Signals appear as bright vertical lines.

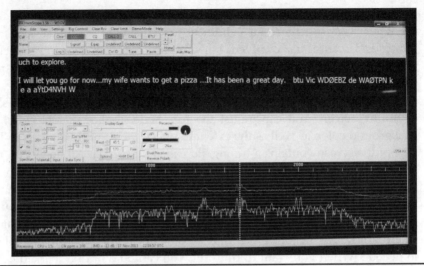

FIGURE 3-6 The HamScope program working in BPSK31 with a spectrum tuning aid. The upper trace shows a time-averaged response; the lower trace shows an instantaneous response.

graph, and instantaneous amplitude versus frequency as the lower graph. Signals appear as vertical "bumps" called *pips*. When you want to decode a signal, you move a pointer (which looks like a tall, skinny human figure between two vertical lines) to its pip, and then left-click your mouse to make a bold, vertical, dashed line move there. If you like, you can enlist the assistance of *automatic frequency control* (AFC) that helps the program get on the correct frequency for optimum decoding. When you want to transmit, you hit a button on the control bar that says "Rx," and it will go over to "Tx." To go back to receive mode, hit the "Tx" button.

As you familiarize yourself with DigiPan or HamScope or any other program that you decide on, I recommend the COEISASWH (click on everything in sight and see what happens) approach. You can try out the "Help" menu, but it might not work on all computers. In any case, you had better prepare yourself for a learning curve when you start working with communication programs like these. For example, you'll need to select the correct COM port for your computer audio before you can get the program to work at all, and you'll have to play around with the audio levels in order to optimize everything in transmit mode and make sure that you put out a decent signal.

When adjusting your equipment to send BPSK31, or any other digital mode, using audio tones into the microphone input of a radio, you run the risk of *intermodulation distortion* (IMD or "intermod") that will produce *splatter*, or signal artifacts that come out near, but on either side of, your operating frequency. You can minimize IMD by adjusting the interface gain, computer audio output, and/or microphone gain controls in your radio so that the audio level remains slightly below the threshold at which your radio's *automatic level control* (ALC) kicks in. Most transceivers will show you, on the front-panel display, when that happens.

Here's a Trick!

Nearly all PSK31 programs and interface units generate audio tones in the range of approximately 300 Hz to 3000 Hz, and send them to the microphone input of a transceiver in SSB mode. While the audio quality of these devices is decent, it's not perfect. Some harmonics occur, and if they get into your SSB transmitting passband, they'll show up as secondary signals on the air. To avoid this problem, make sure that when you transmit a PSK31 signal, you set the audio tones to a frequency of at least 2000 Hz so that all the audio harmonics fall far outside your transmitter filter passband and get suppressed. Figure 3-7 shows an example of the effect. It's a HamScope spectrum-mode display of a W1AW bulletin in BPSK31, tuned to a fundamental audio frequency of 689 Hz after passing through a 250-Hz-wide CW filter in my receiver. You can see the third harmonic at 2067 Hz as a prominent pip. The audio circuits in my receiver, my computer, or both must have spawned it because I know that it could not have come from W1AW! If my receiver's audio circuits can play this trick on me, your transmitter's audio circuits can play it on you. Don't let such garbage go over the air with your call sign on it!

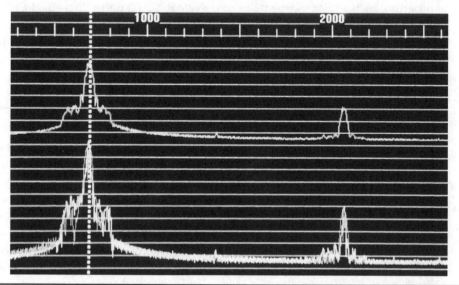

FIGURE 3-7 Harmonic distortion commonly occurs in the audio stages of radios and computers. Don't let it get you in trouble when you transmit digital signals using audio tones at the microphone input of your radio!

Multiple-Frequency-Shift Keying

As you tune around the HF band segments in which digital signals other than CW prevail, you'll hear all sorts of tweets and chirps and tone bursts, some of which sound rather musical, others of which sound like exotic birds or insects. After using RTTY and BPSK31 for a little while, you'll learn to recognize those modes by the way they strike your ears. But what about some of the other signals you hear?

What Is MFSK?

One of the most interesting and promising new digital modes is *multiple-frequency-shift keying* (MFSK). In practice, it's a form of FSK similar to ordinary RTTY, but instead of using two different well-defined carrier frequencies, MFSK uses several different well-defined carrier frequencies. The two primary "flavors" are MFSK16 with 16 frequencies, and MFSK8 with eight frequencies. Radio hams use MFSK16 more often. When you tune in a signal, it comes out of your speaker or headset as a rapidly fluctuating sequence of audio tones.

Error Correction

One of the outstanding features of MFSK16 is *forward error correction* (FEC), a form of *data redundancy* in which the transmitting station sends every data bit more than once. That way, the receiving station can be "pretty sure" that it got each bit correct. This process slows down the transmission speed and causes the data to arrive in spurts or bursts at the receiving end. When you tune in an MFSK signal using a program, such as HamScope that can decode it, you'll notice this effect right away. You'll see a few characters arrive all at once on the screen, followed by a pause, followed by a few more characters all at once, and then another pause, and so on. At full speed, a ham radio MFSK signal moves along at about 42 wpm, averaged over time.

Tech Tidbit

Some error correction methods employ *handshaking*, in which the receiving station sends bit sequences called *packets* at frequent intervals back to the transmitting station so the transmitting station can check for accuracy and repeat a packet if it's wrong. Handshaking, also known as *automatic repeat request* (ARQ), is the best way to ensure data integrity, but it's agonizingly slow at the bandwidths allowed on the HF ham bands, and it requires some sophisticated programming. The transmitting station must frequently pause so that its own receiver can pick up and check requests from the remote receiver. Similarly, the remote receiving station must frequently transmit the requests. In addition to the programming, ARQ mandates a reliable, fast *transmit/receive* (T/R) *switch* at both ends of the circuit.

Other Features

Another notable feature of MFSK16 is the way the signal modulation keeps going in transmit mode, even when you aren't typing anything on the keyboard. Known as *idle insertion*, it keeps the receiver synchronized with the transmitter during pauses in the transmission, so that noise bursts or stray signals don't throw the receiver off and degrade the performance of the system. And that performance can, I have heard, be spectacular indeed. Although I have not witnessed it personally, I've read that signals will sometimes show up on-screen even though a human operator listening to the receiver audio can't tell that any signal exists. If that's true, then MFSK16 might surpass CW as the mode of last resort in adverse conditions. I've displayed MFSK16 from W1AW bulletin transmissions under conditions so bad that I had trouble at times telling that the signal was still there.

How's Your Luck?

If you spend a lot of time working with MFSK16, you might eventually have the experience that I read about on a blog somewhere: A display that spontaneously starts to show text even when all you can hear is noise. As I mentioned a moment ago, at least one ham has reported that MFSK16 signals can "print" even when the human ear can't hear anything other than receiver hiss and sferics. Searching for this phenomenon might make you feel a little like an astronomer working on SETI (the Search for Extraterrestrial Intelligence). Good luck!

Using MFSK16

When receiving MFSK16 signals, you have to tune them in precisely or else you'll get no results. The first thing you'll want to do is to make the receiving program recognize the signal boundaries at the upper and lower frequency limits. Figure 3-8 shows how this works in HamScope. The photo at A shows the program in spectrum mode, and the photo at B shows the waterfall mode. You'll see a pair of dashed lines indicating the standard points for the MFSK16 frequency boundaries. (I prefer the spectrum mode with this program because the dashed lines extend all the way up and down the display; in the waterfall mode they don't.) You'll want to move the dashed lines around with your mouse until they coincide as nearly as possible with the signal boundaries. At some point, you should start to see text on the display.

The MFSK mode is sideband sensitive, so if you've got your radio set to receive signals in the LSB mode and the operator at the other end is sending the signal in the USB mode, you'll have to reverse the *sense* or *polarity* of your reception. An "upside down" signal won't "print." In most MFSK16 programs, you can do that by either hitting a reversal button, or else switching your radio over to the opposite sideband. In HamScope, the button bears the label "Reverse Polarity." If you use a

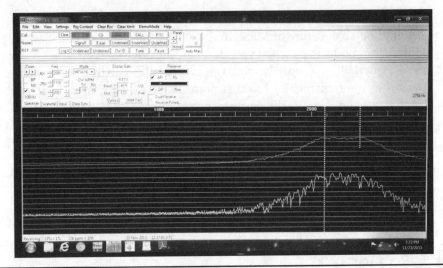

FIGURE 3-8A   HamScope display of the MFSK tuning method called the spectrum mode. To tune a signal in, you adjust the positions of the dashed lines so that they coincide as closely as possible with the "edges" or "skirts" of the signal.

different program, you might see it as "Rev" or "Reverse." Just poke around until you find it!

You'll probably have to exercise your powers of patience when tuning in MFSK16 signals. Not only can the tuning get quite critical in regards to setting the frequency, but also you can't always easily see the signal in a graphical tuning aid display, so you end up guessing at where the dashed lines (or other tuning boundary markers) should go.

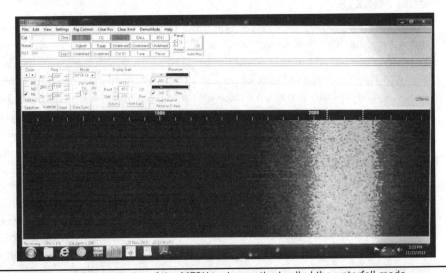

FIGURE 3-8B   HamScope display of the MFSK tuning method called the waterfall mode.

What's That?

If you tune your radio to frequencies a few kilohertz above the "standard" PSK31 bands on HF, you'll hear signals that sound like slow versions of MFSK or PSK—and that's exactly what they are, with a special time-synchronization feature and slow speed to drastically improve the signal-to-noise ratio, even over CW, by as much as 10 dB according to its proponents. This communications scheme has several "flavors" that go by arcane names including WSJT, JT65, JT6M, and FSK441. As of this writing, you can read about it if you point your Web browser to

http://hflink.com/jt65/

If you want to see a lot of technical details and you have an interest in getting on the air with one of these modes, hit the "specifications" link in the page at the above URL.

Amateur Teleprinting over Radio

In ham radio practice, the first mode to take advantage of ARQ handshaking for error correction became known as *amateur teleprinting over radio* (AMTOR). Obviously, all wireless text modes qualify as "teleprinting over radio," including RTTY, PSK, and MFSK. But AMTOR differs from those other modes because it involves continuous exchanges of data in both directions, with frequent switchovers (once or twice per second), once contact has been established. A few pioneering hams started using AMTOR in the early 1980s after having adapted it from the commercial mode called *simplex teleprinting over radio* (SITOR). It has an outstanding advantage over RTTY, PSK, and MFSK: You can maintain almost error-free, albeit sluggish, communications even when lots of fading, interference, or noise occur, if you're willing to take the hardware and programming steps necessary to work in the ARQ mode.

One-Way Error Correction

When a station sends a CQ ("calling anyone who wants to answer") in search of a contact, AMTOR works pretty much like ordinary RTTY using FSK, although AMTOR employs a more sophisticated code that allows for uppercase and lowercase letters as well as numerous punctuation marks and symbols. The AMTOR code has seven bits for a total of 128 (2^7) possible combinations, as opposed to the 32 (2^5) possible combinations in the five-bit Baudot code. An AMTOR CQ sounds just about the same, when it comes from your loudspeaker or headset, as an ordinary RTTY signal does. Error correction is incorporated in the form of FEC, similar to the method used in MFSK.

Two-Way Error Correction

When a two-way contact (or QSO) is established between different stations, AMTOR switches over to the ARQ mode of error correction, which some engineers call *backward error correction* (BEC). The transmitting station sends three characters, and then pauses and waits for a confirmation from the receiving station. If the receiving station gives the "okay," then the transmitting station sends the next three characters. If the receiving station says nothing or says "not okay," then the transmitting station repeats the three characters previously sent.

When you listen to a QSO between two stations using AMTOR, assuming that you can hear them both, you'll recognize it right away by its "tweet, tweet, tweet" sound. In many cases, one station will come in better than the other, so the "tweets" will alternate between strong and weak. Sometimes you'll hear only one of the stations, so the "tweets" will alternate with brief periods of silence. The only problem these days is finding an AMTOR signal to check out! The mode seems to have fallen out of favor, perhaps because of the immense success and popularity of PSK31, and increasing technical interest in MFSK and WSJT modes.

How Fast Is It?

As you can imagine, the real-world speed of an AMTOR transmission depends on communications conditions between the two stations. If there's not much interference, fading, or noise, then the signal cruises along at maximum speed because the transmitting station must send each group of three characters only once. However, if conditions are not so good, the transmitting station has to repeat some sets of characters. The worse the conditions grow, the more the sets of characters have to be repeated, and the slower, in practice, the communications get. Under poor conditions, this process makes for tedious contact, but with fewer errors than in modes, such as conventional RTTY, that don't use ARQ. If conditions get prohibitive, then in effect the speed goes down to zero because acknowledgments never get received.

How Does It "Know"?

Do you wonder how the receiver in an AMTOR contact "knows" whether or not an error has occurred? Actually, it doesn't have 100-percent certainty, but it gets a good clue by comparing the number of mark bits to the number of space bits in each seven-bit code character. In the AMTOR code, every legitimate character has four mark bits and three space bits. The sequence can change, but the ratio is always 4:3. If the receiver "sees" a character that does not adhere to the 4:3 mark-to-space ratio rule, then it can conclude that an error has occurred. Once in a while, however, an error will occur and the resulting signal will still adhere to the 4:3 rule; in that case, you'll get a misprint.

What Happened to It?

Despite its interesting nature and technical advantages, you won't hear AMTOR very much on the air these days. The ARRL official station W1AW no longer sends its non-CW digital text bulletins in AMTOR; as of this writing, they use MFSK16, PSK31, and conventional Baudot RTTY in a rotating sequence. Some innovations "take hold" and others don't; evidently AMTOR didn't.

Packet Radio

A *packet* is a block of digital data sent from a computer at a *source* station to another computer at a *destination* station. The destination-station operator doesn't have to directly attend to the equipment in order for a packet to be received; the computer can store it. For this reason, *packet communications* is a form of *time-shifting communications*. It's a lot like *electronic mail* (e-mail) in that respect.

Networks

Wireless packet networks are getting larger, easier to access, and more sophisticated all the time. The data speed (thousands of baud) exceeds that of other popular ham radio digital modes, allowing for the transmission of fairly long messages. Packet communications is self-correcting. The destination equipment detects discrepancies in received data by means of handshaking, similar to the method used in AMTOR. In an ideal packet network, you can send a message to any destination, anywhere in the world, and have confidence that the network will automatically get it there within a few minutes or, at most, a few hours.

Amateur-Radio packet communications is called *packet radio*. Instead of connecting the computer to the commercial Internet by means of a cable, wireless, or satellite modem, you interface the computer with a ham radio transceiver using a *terminal node controller* (TNC), which acts as a special-purpose modem with built-in memory. Packet radio is, in effect, computer communications by wireless means, and in ham packet radio, that communications can take place (if desired) entirely apart from the Internet. If the utility and commercial infrastructures go down, packet radio will still work to some extent if you're willing to use HF for long-haul links.

The Terminal Node Controller

A TNC assembles packets that are composed on, and stored by, the computer, and then converts the packets into a form suitable for transmission by radio using *digital-to-analog* (D/A) *conversion*. The TNC also takes packets from the radio receiver, and puts them in the proper format for display and/or storage on the computer; that's *analog-to-digital* (A/D) *conversion*. If a TNC has its own memory, it can store an incoming message indefinitely, and send it out again at any time. The computer can also store the message. In effect, therefore, a well-equipped amateur packet radio station can function as a communications *repeater*. Because the packets themselves exist in digital form, repeaters of this sort have gotten the nickname *digipeater*. Each digipeater in a packet radio network constitutes a *node*.

A Packet-Equipped Ham Station

Figure 3-9A shows a simple ham packet radio station. The personal computer is equipped with Internet access (such as a cable, wireless, or satellite modem) as well

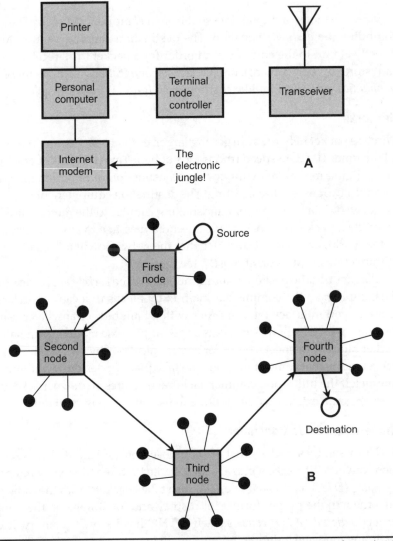

FIGURE 3-9   At A, a ham station equipped for packet radio. At B, the passage of a packet through nodes in a wireless communications circuit. Black dots represent end users.

as a TNC, so messages can be sent and received by means of conventional online services as well as with the transceiver. With this arrangement, you can receive a packet from the radio, store it in your computer, and send it to someone at a specific e-mail address on the Internet or to another ham with a packet-equipped station, either directly or through one or more digipeaters.

When All Else Fails

Hams can exchange packet radio messages among themselves and with their own digipeaters, radios, and computers, without involving the Internet at any point whatsoever. In that sense, a collection of ham radio packet stations and digipeaters can serve as an "alternative Internet." That capability might prove handy someday if the regular Internet suffers a massive outage, either as a result of natural events or as a result of human sabotage! Ham radio will then get a chance to show its value for public service, as it has done during and after many past disasters.

Nodes and Links

Figure 3-9B shows a generic example of how a packet radio message gets routed. The black dots represent *subscribers* (end users) with packet communications capability. The four rectangles represent *local nodes*, which can exist at geographically diverse locations. Each local node serves several stations in its immediate vicinity by means of short-range *links* on the amateur VHF, UHF, and microwave bands. The nodes can be interconnected by terrestrial radio links if they're relatively near each other. If the nodes are widely separated, satellite or HF links are used. Unfortunately, HF packet communications is rather inefficient and slow because of variable propagation conditions. Nevertheless, it can serve as an option when everything else fails.

In a packet radio network, the source station needs to know a specific *address* that identifies the local node of the destination station, as well as a specific address for the destination station itself, in order to send the message and ensure that it gets where it's meant to go. The transfer then takes place automatically, in much the same way as data goes over online computer networks, except that some of the link goes over ham radio media, which by law can't be commercialized. If you aren't a radio ham but have occasion to "employ" a ham radio operator to get a message to a loved one in a disaster area someday, the radio ham won't (and legally can't) accept any remuneration for the service.

Bulletin Boards

In Internet practice, a *bulletin-board system* (BBS) is a "forum" for stored messages, accessible by means of a computer. A BBS lets you use a modem to connect your computer to the Internet—or to a ham packet radio network, in which case you have a *packet radio bulletin-board system* (abbreviated PBBS)—and leave messages for other users, just as you would do with paper and thumb tacks on an old-fashioned bulletin board. With a TNC connecting a computer to a radio transceiver, you can access numerous PBBSs. You can leave messages for, and read messages from, other hams or general Internet users at any time. You can leave messages for as long as you want, and then delete them when you want them gone from the system.

A PBBS can function even when the destination operators are not physically present to look after their stations. That's why you'll sometimes observe a significant delay in transferring messages. The operator of the destination station must check the PBBS before he or she can get the message. A delay might also occur if the network has difficulty routing the packets. This can happen because the network gets overloaded with too many users at a single time, or part of it gets physically compromised (by a hurricane or wildfire for example). If part of the link goes over the HF bands, the likelihood of long delays increases.

Why So Many Modes?

You might ask, "Why should hams spend time working on new digital communications modes when perfectly good ones, like CW and RTTY, already exist?" As an old timer in this hobby, I can answer that question in two parts: (1) Technically inclined radio hams have an inborn compulsion to dream up new modes and get them to work; and (2) these people enjoy an endless quest for minimizing error rates and maximizing digital communications efficiency. Combined, these two ingredients form the perfect framework for technological progress. Some of the most significant innovations in communications history have arisen from the activities of radio amateurs "just having fun."

Single Sideband

By far the most popular HF mode, single sideband (SSB, also called "phone," a throwback to the days when voice communications was called "radiotelephone") evolved from the older, less efficient AM technology. Basically, an SSB signal is an AM signal with the carrier and one sideband removed, and then filtered so that the bandwidth is limited to a little less than 3 kHz. In most radios these days, the audio input is filtered for a passband that ranges from 300 Hz to 3 kHz, so the practical bandwidth is 2.7 kHz. Then, in addition to the audio filtering, the RF stages of the transmitter incorporate a second bandpass filter in the IF chain.

Which Sideband Should You Use?

In amateur HF activity, most hams use the lower sideband (LSB) on the bands below 10 MHz, and the upper sideband (USB) on the bands above 10 MHz. There's no technical advantage to using either sideband on any frequency; these standards have evolved as a matter of convention.

If you have your radio set to receive USB, you'll find it impossible to tune in an LSB signal so that you can understand what the operator says. All you'll hear is "monkey chatter," the result of the higher voice frequency components coming in at low frequencies and vice versa. You'll observe the same effect if you set your radio to receive LSB and you try to bring in a USB signal.

On the VHF and UHF ham bands, a smaller proportion of the total operators use SSB than is the case on the HF bands. Above 50 MHz, most voice communications activity takes place with FM, often using repeaters that receive and retransmit signals to extend the range of communications. However, those operators who do use SSB on VHF or UHF (mainly on the 50-MHz band) almost always employ USB.

Tip

If you're having an SSB contact and you want to gain a certain measure of privacy, switch to the sideband that goes against convention. That way, people casually tuning through the band will not likely understand your transmissions unless they deliberately switch sidebands in order to pay attention to you. But remember that ham radio is a public communications medium; nothing, technologically or legally, can prevent someone with a good receiver from eavesdropping on your on-the-air rantings.

VOX or PTT?

The abbreviation PTT stands for *push to talk*. That's the simplest and most common way to operate SSB. When you want to transmit, you press and hold down a button, usually a momentary-contact switch on the side of the microphone. In most radios, you can squeeze the microphone to close the switch and talk.

The acronym VOX stands for *voice-operated transmission* (hams often abbreviate the word "transmission" as TX or X). In SSB mode, VOX allows you to actuate your transmitter using the electrical voice impulses from a microphone or audio amplifier. It can do this with either a relay or an electronic switch.

The VOX mode eliminates the need for throwing a switch or pressing a button to change from receive mode to transmit mode. This feature can prove especially useful in mobile operation, when you need both hands for such things as operating other controls (your steering wheel, for example). You can disable the VOX and use PTT instead if you prefer that method.

In VOX operation, the transmitter is actuated within a few milliseconds after you speak into the microphone. The transmitter remains on for a short time after you stop speaking; the delay prevents your radio from "tripping out" during short pauses, a phenomenon that most people find exasperating. Radios equipped with VOX allow you to adjust the delay time between transmit and receive modes after you stop speaking.

Did You Know?

In a radio with VOX capability, an *anti-VOX* system uses *negative feedback* to keep the VOX from kicking in when your microphone picks up audio from your receiver's speaker. You can adjust the anti-VOX sensitivity so that it will keep your VOX circuit from "chasing its own tail," and yet still allow you to transmit by speaking into the microphone. Of course, you can circumvent the need for anti-VOX by using a headset instead of a speaker for reception. Even better, try a headset with a built-in microphone, like telephone switchboard operators used in days long gone by!

Automatic Level Control (ALC)

In an SSB transmitter, *automatic level control* (ALC) prevents excessive modulation (or *overmodulation*) while allowing ample microphone gain. Overmodulation causes a phenomenon known as *splatter*, which, if not filtered out somewhere in the IF or RF chain of the transmitter, greatly increases the emitted signal bandwidth and can cause interference to stations on nearby frequencies.

Figure 3-10A shows an amplitude-versus-time graph of a fully, but properly, modulated SSB signal with a sine-wave audio input. If overmodulation occurs, *peak clipping* (Fig. 3-10B) can result, producing sideband elements at frequencies outside the normal passband, as shown in Fig. 3-11. That's splatter!

In a receiver, splatter sounds like a crackling noise at frequencies above and below the actual signal. Such emissions not only annoy other operators, but they can also generate disruptive interference. Splatter also violates FCC regulations because it causes an SSB signal to exceed its legally allowed bandwidth.

Speech Compression and Clipping

Speech compression is a method of increasing the average power in a voice signal, without increasing the peak power. A speech compression circuit operates in much the same way as an ALC circuit does. But speech compression carries the process farther than ordinary ALC.

While ALC is intended only to prevent overmodulation, speech compression employs additional amplification of the low-level components of a voice in an effort to get more "bang for the buck" out of the signal. The result is reduced dynamic range (difference between the loudest and the softest voice input levels),

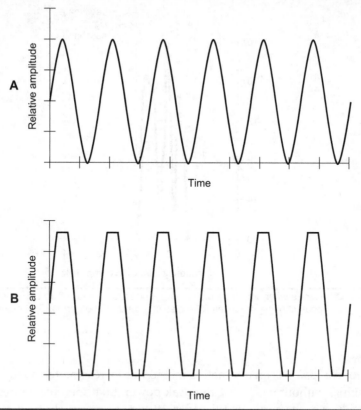

FIGURE 3-10 At **A**, an amplitude-versus-time graph of a fully modulated SSB signal with a sine-wave input at the microphone, free of peak clipping. At **B**, an amplitude-versus-time graph of an SSB signal with the same sine-wave input but suffering from peak clipping, which can produce splatter.

but the intelligibility can improve, especially under marginal communications conditions, because the signal carries extra "heft." The peak power doesn't increase compared to a signal without compression, but the average power does.

Speech compression, also called *amplified automatic level control* (AALC), is usually carried out in the microphone-amplifier circuits of a transmitter. But it can also be done in the RF stages. Accordingly, engineers may speak of either audio speech compression or RF speech compression, also known as *envelope compression*.

When speech compression is used in a transmitter, the transmitter output should be monitored with a spectrum analyzer or panoramic receiver to ascertain that splatter does not take place. Also, precautions should be taken to ensure that the average-power increase will not put too much strain on the final amplifier stage of the transmitter.

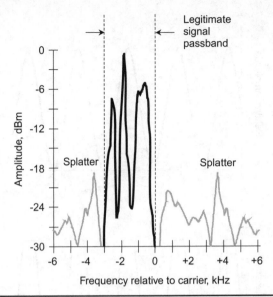

FIGURE 3-11   Here's what splatter looks like on a spectral display of an LSB voice signal. The peak clipping produces RF emissions outside the normal SSB passband.

Speech clipping is an alternative method of increasing the average power of an SSB signal without increasing the peak power. Like speech compression, it can be done either in the audio or RF stages. If done properly, clipping can improve the intelligibility of an SSB signal in the same way as compression does, but clipping generates splatter as a matter of course! Whenever speech clipping is used, therefore, a high-quality, 3-kHz-wide bandpass filter must follow the clipping circuit to keep the splatter off the air.

Frequency Modulation

Radio hams use FM almost entirely on the VHF and UHF bands, that is, 6 meters and shorter wavelengths (50 MHz and above). Most operation these days takes place through repeaters, which receive signals on a certain frequency and then retransmit them on slightly different frequencies.

Bandwidth

When you frequency-modulate a carrier wave, you get sidebands above and below the carrier frequency, just as you do with AM. But there's a difference in the way the sidebands are arranged, frequency-wise and bandwidth-wise. As you increase the deviation, additional sidebands appear, farther away from the carrier frequency than the ones that you would see with AM. The number and relative amplitudes of these sidebands is a complex mathematical function of the deviation, and also of the highest audio frequency that goes into the transmitter.

If an FM signal's bandwidth does not exceed that of a standard AM signal having the same waveform, engineers call it *narrowband FM* (NBFM). For voice communications, the maximum bandwidth for NBFM is about 6 kHz, or 3 kHz above and below the carrier. Hams do most FM operation on their VHF and UHF bands in the narrowband mode.

If an FM signal has a bandwidth greater than that of a standard AM signal with the same modulating waveform, then it's called *wideband FM* or WBFM. Hi-fi stereo broadcasting in the band from 88 MHz to 108 MHz provides a good example. The fidelity improves as the bandwidth increases (just as it does in any other mode), but in ham radio practice, intelligibility surpasses fidelity in importance. The FCC forbids ham radio operators from transmitting music, anyway!

Noise Immunity, Sort Of

For voice communications, FM offers better immunity to static and impulse noise than AM does because these types of noise are, in effect, amplitude-modulated. Receivers for FM can be made unresponsive to changes in signal amplitude by means of limiter circuits. Nevertheless, in practice, weak FM signals that fall far below the limiter threshold are affected by static or impulse noise because the signals aren't strong enough to make the limiting circuits "kick in."

Repeaters

In Amateur Radio practice, repeaters are used mostly on the VHF and UHF bands. Since about 1970, repeaters, often funded and maintained by local ham radio clubs, have revolutionized the nature of communications on these bands, especially 144 MHz (2 meters) and 440 MHz (70 centimeters).

A repeater receives a signal on a particular assigned frequency, and retransmits it at the same time, usually in the same ham band but on a different assigned frequency. A good repeater will increase the range between mobile and portable stations compared with direct, or *simplex*, communications. Any two transceivers that are within the repeater range can communicate with each other, even if they can't do it on simplex.

Ideally, a repeater should be located in a high place, such as on a mountain or tall building (Fig. 3-12A). The repeater transmitter can, and often does, have much higher RF power output than a ham radio operator's own transmitter. An omnidirectional gain antenna enhances the receiving and transmitting range of the repeater. The radius of reliable repeater coverage can exceed 100 kilometers when all these factors are optimized. Therefore, two hams using mobile equipment might have a QSO (conversation) over a distance of 200 kilometers or more. Without the repeater, the maximum communications range would be about 50 kilometers over flat terrain, and less than that over irregular terrain.

Tip

If you plan to build and design a repeater, contact the American Radio Relay League (ARRL) for advice and direction. Their website, once again, is

www.arrl.org

You'll have to make yourself aware of the regulations and procedures involved with repeater installation, operation, and maintenance. The ARRL can tell you everything that you need to know.

Consider This!

I recommend that you build and provide an emergency backup power system for your repeater! You might even consider taking it entirely off the grid and running it with solar or wind power, or a hybrid system that takes advantage of both. Repeaters, even large ones, don't consume a lot of electrical energy, so off-grid solutions are practical. Then your system will be ready to serve in case of a disaster that knocks out the regular electrical utility.

A basic repeater includes a receiver, a transmitter, an antenna, and an isolating network called a *duplexer* (Fig. 3-12B). The incoming signal is picked up by the antenna and fed to the receiver. The demodulated output of the receiver goes directly into the audio input of the transmitter. The transmitter retransmits the signal at the same time, using the same antenna as the receiver.

In a repeater, the transmitter signal must differ in frequency from the received signal by at least a certain amount, called the *offset* or *split*, which varies depending on the band. Otherwise the transmitter will overload the receiver. The split generally gets larger as the repeater's operating frequency increases. On 2 meters, 600 kHz is the most common split. On 70 centimeters, the split is usually 5.00 MHz.

Most modern transceivers have a provision for programming the split frequency in case you encounter a repeater that has a nonstandard split. If you don't deliberately program the split in, the radio will operate with a default split representing the most commonly used value for the frequency to which the receiver is tuned.

Tone Squelching

In recent years, more and more radio clubs have equipped their repeaters with *tone squelching*. In this scheme, the repeater's receiver incorporates a silencing circuit that stays closed unless an incoming signal is modulated with a tone of a certain frequency and/or duration. People who want to use the repeater must know the tone frequency and program their radios to put it out over the air, or the repeater won't accept their signals. A repeater that uses tone squelching to regulate access is called a *closed repeater*. If you can access a repeater using any radio without tone squelching, then you have an *open repeater*.

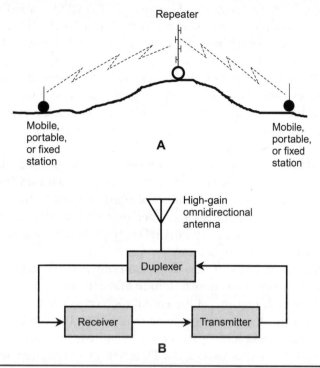

FIGURE 3-12 At **A**, the principle of repeater operation. At **B**, a simplified block diagram of a ham radio repeater.

Images and Video

Radio amateurs have been sending and receiving images, both fixed and moving, for decades. The most common ham radio image modes are television and facsimile (FAX). Television takes two forms: analog fast scan and slow scan. Most hams stick to grayscale images and videos to conserve spectrum space and get the best possible image resolution (detail). However, color images have been successfully sent and received on all the bands on which image transmission is allowed.

Fast-Scan Television

Amateur fast-scan television (FSTV) signals are just like old-fashioned analog TV broadcast signals, except that hams use less transmitter power than broadcasters did! An analog FSTV system provides 30 images, or *frames*, per second. Each frame has 525 lines in the screen rendition or *raster*. The lines run horizontally across the screen, which has a horizontal-to-vertical dimensional ratio, or *aspect ratio*, of 4:3.

Each line contains shades of brightness in a grayscale system, and shades of brightness and hue in a color system. In FSTV broadcasting, the image is generally sent as an AM signal, and the sound goes out as an FM signal. A standard FSTV channel takes up 6 MHz of spectrum space, or 1000 times the bandwidth of an AM voice signal! For this reason, all amateur FSTV activity takes place on the UHF bands at 70 centimeters and above. On the lower bands, FSTV signals are simply "too big." For example, the entire 2-meter ham band is only 4 MHz wide, not even enough for one FSTV signal.

An FSTV transmitter comprises a camera, an oscillator, an amplitude modulator, and a series of amplifiers for the video signal. The audio system consists of an input device such as a microphone, an oscillator, a frequency modulator, and a feed system that couples the RF output into the video amplifier chain. It also has an antenna output, of course!

An old-fashioned analog FSTV receiver, which can be used without modification to receive ham FSTV signals, has input terminals with an impedance of either 75 ohms unbalanced or 300 ohms balanced, a tunable front end, an oscillator and mixer, a set of IF amplifiers, a video demodulator, an audio demodulator and amplifier chain, a cathode-ray tube (CRT) or "picture tube" with associated peripheral circuitry, and a speaker or headset.

In order for an FSTV picture to appear normal, the transmitter and the receiver must maintain exact synchronization. To that end, the transmitter generates pulses at the end of each line and at the end of each complete frame in the raster. These pulses get sent along with the video signal. In the receiver, the demodulator recovers these synchronizing (sync) pulses and sends them to the display. In a CRT display, such as the type used in old TV receivers, the electron beam inside the tube moves in exact synchronization with the scanning in the camera.

An Old Timer Remembers

If the frame sync in a "vintage TV set" gets upset for any reason, the picture will *roll*, or move vertically from frame to frame. If the line sync is not right, the picture will *tear*, or get ripped up horizontally so that it looks like … well, like a ripped up image! Most old analog FSTV receivers have controls that allow adjustment of the frame and line sync, usually called "vertical hold" and "horizontal hold." You have to set them in order to get a good display.

Figure 3-13 is a graph of a single line in an FSTV video signal, representing 1/525 of a frame. The highest signal level corresponds to the darkest shade (ideally black), and the lowest signal level corresponds to the lightest shade (ideally white). Therefore, the FSTV signal gets modulated "negatively." This convention allows for *retracing* (moving from the end of one line to the beginning of the next) to synchronize between the transmitter and receiver. A well-defined, strong *blanking pulse* tells the receiver when to retrace, and it also shuts off the beam while the receiver display retraces by effectively turning the screen as black as it can get during that brief period of time.

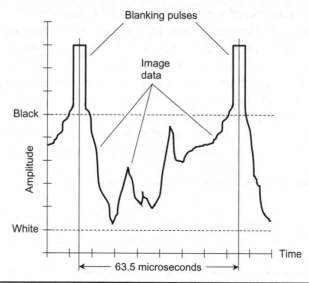

FIGURE 3-13   A single line in an analog FSTV video frame as it would appear on a lab oscilloscope.

Slow-Scan Television

A fast-scan television signal of the sort used in television broadcast takes 6 MHz of spectrum space because of the high resolution of the image, along with the rapid changes necessary to produce full-motion video. With some compromise, however, you can send a video image in a much narrower band. *Slow-scan television* (SSTV) accomplishes this feat in part by reducing the rate at which pictures get generated and sent. An SSTV signal needs only 3 kHz of spectrum space, equivalent to SSB in bandwidth, so SSTV is practical on HF.

A fast-scan amateur television signal has 30 complete picture frames per second. A slow-scan signal sends a maximum of one frame every 8 seconds—240 times slower, allowing a dramatic reduction in bandwidth. (Some SSTV signals go out even slower than that.) A typical SSTV signal has 120 lines per frame, compared with 525 lines per frame in FSTV. This reduction further reduces the bandwidth of SSTV.

To get an SSTV signal, you can input audio data into the microphone jack of an SSB transmitter. A tone of 1500 Hz corresponds to black; 2300 Hz gives white. Intermediate audio frequencies produce shades of gray. Synchronization signals are sent at 1200 Hz as short bursts lasting 30 milliseconds for vertical synchronization and 5 milliseconds for horizontal sync.

You can transmit SSTV images and voice transmissions at the same time by sending the pictures on one sideband and your voice on the other sideband. Then you'll be able to carry on a continuous voice narration with fixed images every 8 seconds, something like a slide show.

If you want to use an analog television set to display SSTV images, you'll need a scan converter. There are fewer scan lines in SSTV than in the TV receiver; the converter adjusts for this difference, yielding a picture almost as detailed (except for the lack of motion) as a TV broadcast signal under good band conditions.

An SSTV station without a computer needs a transceiver with SSB capability, a scan converter, a TV set, and a camera. Scan converters and cameras are commercially available from many ham manufacturers and dealers. Alternatively, you can use a computer, equipped with a dedicated program for encoding and decoding SSTV, along with a webcam or the computer's built-in camera, if it has one.

Tip

Hams do their SSTV operating in the SSB portions of the HF bands. You'll instantly recognize an SSTV signal by its sound, once you've heard a few of them!

Did You Know?

Although amateur SSTV is done in grayscale for the most part, a few hams have experimented with color SSTV. As you might guess, color SSTV signals are more subject to the whims of noise and interference than are grayscale signals. The transmission rate is usually a lot slower, too, with a complete frame taking upwards of a full minute to render.

Facsimile

Radio hams can send *facsimile* (FAX) images by means of narrowband transmissions, in a manner very much like the way it works over telephone channels. The bandwidth of a FAX signal is the same as that of an SSB signal.

Amateur FAX transmissions have resolution comparable to that of any commercial transmission. However, ham FAX is done by means of radio, often at HF, where the ionosphere introduces phase modulation and fading, and where atmospheric and humanmade noise and interference commonly occur. These factors degrade the image quality. The problem is less severe or frequent at VHF and UHF.

To send a fax from a hardcopy, you place one page, containing text and/or images, into an *optical scanner* that converts the hard copy into a series of binary digital pulses, that is, *high* and *low* signals (also called 1 and 0). The output of the scanner goes to a modem that converts the binary digital pulses into a signal suitable for transmission over your radio.

At the destination or receiving station, the analog signals from the telephone line or radio get converted back into digital pulses like those produced by the optical scanner at the source or transmitting station. These pulses go to a printing device or computer. A standard fax has a printer similar to a photocopier or laser printer. In fact, a rendered hardcopy fax looks like a photocopy. The image resolution of a high-quality fax is good enough to allow reproduction of most photographs.

A ham FAX transmission requires a few minutes to send. The more time that you allow for sending the image, the better the resolution you'll get. The image goes out in lines, similar to the way in which a television raster is scanned. In ham FAX practice, 120 or 240 lines per minute (two or four lines per second) are scanned. The modulation method can be either positive or negative. In positive modulation, the instantaneous signal level is directly proportional to the brightness at any given point on the image. In negative modulation, the instantaneous signal level is inversely proportional to the brightness at any given point on the image.

Some hams have used surplus commercial FAX apparatus to build their FAX stations. If you want to do that, you'll probably have to modify the equipment to work with your radio, a challenge to the ham who likes to tinker and perfect station equipment. Most commercial units are designed to work with AM, but they can be converted to work with FM. Some hams like to build all of their FAX equipment from scratch. Others prefer to download computer programs and let their cyber hardware take care of it all!

CHAPTER 4

Ham Radio Licenses and Frequencies

Amateur Radio operators enjoy operating privileges in numerous slices of the radio spectrum. Let's take a look at the ham radio bands, learn how they work on the air, and outline the mode and power restrictions imposed by the FCC. The information given here is current as of February 2014 as I write this chapter. But things keep changing for radio hams! I recommend that you periodically visit the ARRL website for updates, including possible new bands or adjustments to existing bands.

Quick Look

Table 4-1 offers an at-a-glance overview of the ham bands available to operators in the United States. That's a lot of bands, but as you'll see a little later in this chapter, it's not a very big chunk of the spectrum in absolute terms! Where will you find your "niche" in this realm?

Today's License Classes

If you don't have a ham radio license but want to get into this hobby, you'll have a choice of three license levels or "classes." Most people start with a so-called *Technician Class license*, which conveys limited privileges, primarily using voice modes. More serious hams go for the *General Class license* or the *Amateur Extra Class license*. You must take written tests to get the licenses, but you don't have to demonstrate any knowledge of, or proficiency in, the Morse code as you had to do in years gone by. The FCC eliminated that requirement after countries throughout the world agreed that "the code" no longer represented an essential communications skill.

Technician

Most new hams hold Technician level licenses. In order to get one, you must take a 35-question test and get at least 74 percent of the answers correct, or 26 "hits."

TABLE 4-1   Amateur Radio bands in the United States. This data is effective as of 2014. For information about possible changes, visit www.arrl.org.

Frequency Range	Wavelength Designator
1.800 MHz to 2.000 MHz*	160 meters
3.500 MHz to 4.000 MHz	80 meters
5.3305 MHz to 5.4035 MHz*#@	60 meters
7.000 MHz to 7.300 MHz*	40 meters
10.100 MHz to 10.150 MHz*@	30 meters
18.068 MHz to 18.168 MHz	17 meters
21.000 MHz to 21.450 MHz	15 meters
24.890 MHz to 24.990 MHz	12 meters
28.000 MHz to 29.700 MHz	10 meters
50.00 MHz to 54.00 MHz	6 meters
144.0 MHz to 148.0 MHz	2 meters
219.0 MHz to 220.0 MHz and 222.0 MHz to 225.0 MHz	1.25 meters
420.0 MHz to 450.0 MHz%	70 centimeters
902.0 MHz to 928.0 MHz%	33 centimeters
1.240 GHz to 1.300 GHz%	23 centimeters
2.300 GHz to 2.310 GHz%	13 centimeters
2.390 GHz to 2.450 GHz%	12 centimeters
3.300 GHz to 3.500 GHz%	87 millimeters
5.650 GHz to 5.925 GHz%	51 millimeters
10.00 GHz to 10.50 GHz%	29 millimeters
24.00 GHz to 24.25 GHz%	12 millimeters
47.00 GHz to 47.20 GHz%	6.4 millimeters
76.00 GHz to 81.00 GHz%	3.8 millimeters
122.25 GHz to 123.00 GHz%	2.4 millimeters
134.0 GHz to 141.0 GHz%	2.2 millimeters
241.0 GHz to 250.0 GHz%	1.2 millimeters
Above 275.0 GHz%	1.1 millimeters and down

*Part or all of the band must be shared with other services.

#Frequencies limited to specific channels. See text for details.

@Maximum RF power output is limited. See text for details.

%Special restrictions may apply, depending on where you live.

Volunteer hams administer the exams. Most local ham radio clubs hold regular examination sessions; if they don't, one of their members will arrange to have a volunteer administer your test by appointment.

The test, as you can guess, deals with some fundamental technical stuff as well as important regulations imposed by the FCC. The Technician Class license test is fairly easy. If you've studied a basic electronics text or two, and if you've carefully gone over the FCC regulations as published in ARRL study guides, you should pass on your first try! Then you can legally transmit with RTTY, data, and voice modes on many, but not all, of the Amateur Radio frequency bands.

Want to Know More?

You'll find details about specific frequency privileges for the various license classes as you read the band-by-band breakdown information in this chapter.

Tip

Call or e-mail the ARRL Headquarters Help Department if you are an aspiring ham radio operator. Let them tell you all about the exam preparation process. Again, their main website is

www.arrl.org

They'll probably offer to sell you some of their specialized license-exam study guides. If they do, accept the offer! If they don't, ask about them. They're all great books. Study them and then go get that license!

General

After you've held a Tech license for a while, you'll probably want to upgrade to the General Class license. Then you'll get nearly all of the frequency privileges that ham radio has to offer, from 160 meters (1.8 MHz) through the highest allocations in the super-high frequency (SHF) and microwave parts of the radio spectrum.

In order to upgrade your license class, you'll have to take another test. The General Class exam goes into more depth than the Tech exam, particularly when it comes to electronics theory and communications practice. But once you have that piece of paper, you can get involved with all the known (and as yet unknown) communications modes.

Extra

The Amateur Extra Class license, often called simply the *Extra*, gives you full ham radio operating privileges on all bands to the extent allowed by the law, which, as I have already warned you, changes fairly often. In order to get this license, you'll need to pass a rather difficult technical exam that has 50 questions, 37 of which you must answer correctly.

If you're serious about ham radio and you really like to tinker with radios and their associated devices, or if you want access to the "prime space" in the HF bands where the best DX (foreign stations in exotic places) activity takes place, or if you simply want to hang out with the "cream of the crop," you should consider getting the Amateur Extra Class license.

To "bone up" for the test, I recommend that you obtain and study the technical guides for the Extra Class license published by the ARRL, as well as my books *Teach Yourself Electricity and Electronics*, *Electricity Demystified*, and *Electronics Demystified* published by McGraw-Hill. They're all available through Amazon and other online bookseller sites. You might even find one in a bricks-and-mortar bookstore. Make certain that you get the latest editions.

Renew!

All Amateur Radio licenses remain valid for 10 years from the date of issue. You can renew your license at the same level without taking the test over again. You have a grace period to renew your license if you allow it to lapse, although you can't transmit with your radio until you get a valid renewed license. But why let it lapse?

Discontinued License Classes

Once in a while you'll encounter a ham who claims to hold a license of some class other than the three described above. What's up with that? Well, the ham radio licensing levels used to be a lot more complicated than they are today. The old classes went by the monikers *Novice*, *Technician Plus*, and *Advanced*. Although the FCC no longer issues new licenses for these classes, original ones remain valid.

The Novice Class license conveys restricted CW privileges on a few of the HF bands, along with some RTTY, data, and SSB privileges on 10 meters. In addition, voice and image mode privileges are available on 1.25 meters at 222 to 225 MHz, and also on 23 centimeters at 1.270 to 1.295 GHz.

The Technician Plus Class license "morphed" into today's Technician Class license. In days gone by, Techs couldn't operate on any of the HF bands; they were confined to 50 MHz and above. Today Techs can use the same CW band segments as Novices can, in addition to RTTY, data, imaging, message forwarding, and other more exotic modes on the VHF, UHF, SHF, and microwave bands.

The Advanced Class license originated many years ago, and then the FCC phased it out for a couple of decades. Later still, around 1970, the FCC brought it back when they implemented a stratified scheme known as *incentive licensing*. Presumably the FCC wanted to motivate as many hams as possible to gain technical knowledge and operating proficiency.

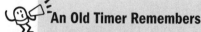

An Old Timer Remembers

I recall the time when I, as a General Class licensee, lost a lot of my operating space; as a new "General," I had just gotten all the privileges ham radio had to offer. I reacted to this change with dismay, and then decided that I might as well upgrade to Advanced and then to Extra a few months after that. I loved ham radio and I wanted all of it! But some hams got furious when they saw nearly half of their privileges on SSB and/or CW taken away. One guy called the plan "invective licensing." I guess people get unhappy when their government takes things away from them.

The Advanced Class license has vanished once more. New Advanced licenses are no longer issued. Some of the privileges taken away during the incentive licensing days have been restored. Advanced Class hams get to operate on a few more frequencies than Generals do, but Advanced operators don't get everything.

Tip

At the time of this writing, I see a fine website that goes into detail about phased-out, as well as currently existing, license levels. Go to

www.hamradiolicenseexam.com/which-exam.htm

As with other government-related matters, a bunch of rules and regulations go along with all this information. Those rules can get confusing. Read through the contents of the website if you hold one of those older license classes (Novice, Technician Plus, or Advanced), and see how the rules might affect you.

160 Meters

The ham 160-meter band extends from 1.800 MHz to 2.000 MHz. That frequency range lies just above the upper limit of the standard AM broadcast band. Actually, it's part of the MF spectrum, not the HF spectrum, which technically starts at 3 MHz and goes up to 30 MHz. Nevertheless, most hams think of 160 meters as an HF band.

Sharing with Other Services

Ham radio operators share the upper half of this band (1.900 MHz to 2.000 MHz) with radiolocation services. Old timers will remember these services as LORAN, an acronym that stands for *LOng RAnge Navigation*. By law, ham radio operators must avoid interfering with these services.

Characteristics

Experienced 160-meter operators know it as a winter nighttime band. Long-distance ionospheric propagation occurs only over paths that lie entirely or mostly on the dark side of the planet. During the daytime, the D layer ionizes to the extent that it prevents 1.8-MHz radio waves from getting to the higher layers where they could otherwise bend back to the surface. When the sun goes down, the D layer vanishes and 160 meters "opens up."

Because waves at 1.8 MHz are so long, a good antenna must be long and high. A full-size dipole on this band measures 79 meters from end to end, and for optimum performance, it must dangle at least 40 meters above the ground. Not many hams have the real estate for anything like that. But smaller antennas can get you contacts; an inductively loaded vertical, mounted on the ground with a good radial system, will provide hours of fun.

Atmospheric noise in the form of sferics ("static") presents a problem on 160 meters during the months when thundershowers commonly occur. For that reason, and also because the hours of darkness don't last as long during those seasons (spring and summer) as they do in the fall and winter, you'll find 160 meters most usable on cold, dark nights in November through February, when, along with a wood fire and a cup of coffee or hot chocolate, ham radio operation completes a great trio!

Allocations by Mode

The 160-meter band isn't broken down into segments by mode. Hams can use CW, SSB, image, RTTY, and data modes on any frequency in the band.

Allocations by License Class

No class-specific subbands exist within the 160-meter band. Hams who hold General, Advanced, or Extra class licenses may use any frequency between 1.800 MHz and 2.000 MHz. The whole band is off-limits to Novice, Technician, and Technician Plus license holders. Some Novices, Techs, and Tech Plus hams find the desire to get 160-meter privileges sufficient motivation to upgrade!

Did You Know?

Because hams who use 160 meters have gained a long-standing reputation for courtesy and respect towards other operators, this part of the spectrum goes by the nickname "the gentlemen's band." For the most part that remains true, but we might do well to remember that some female hams exist! Maybe we should call it "the good-manners band."

80 Meters

The 80-meter band extends from 3.500 MHz to 4.000 MHz. Sometimes the upper half of this band is called 75 meters. But overall it averages 80 meters: In free space, an 80-meter wave has a frequency of 3.7500 MHz, precisely in the middle of the band. It's a large band, considering its place in the spectrum, occupying fully 12.5 percent of the frequency space between DC (0 Hz) and 4.000 MHz. Hams are lucky indeed to have this band. Moreover, no sharing provisions impede hams' full enjoyment of it. If you're a ham, it's all yours!

Characteristics

Like 160 meters, the 80-and-75-meter band works best at night, and better in the fall and winter than in the spring and summer. It's that way for the same reasons: 80-to-75-meter waves propagate better when the D layer lacks ionization, and that's during the hours of darkness. Also, sferics can present a problem in regions of the world where thunderstorms occur, and in most cases, that's springtime and summertime.

While a full-size 80-and-75-meter antenna isn't as imposing as its 160-meter counterpart, a good dipole at a reasonable height still takes up a lot of space. If you have the land and the clearance for it, a 40-meter-long wire, center fed and elevated 20 meters or more above the ground, will serve you well on all the frequencies from 3.500 MHz to 4.000 MHz if you cut it for the center of the band at 3.750 MHz.

Tip

You'll learn later in this book how to determine the optimum lengths and/or heights for various types of antennas. For now, you can remember this simple formula for figuring out the ideal end-to-end electrical length for a half-wave wire dipole in meters: Divide 143 by the frequency in megahertz. You can almost always operate such an antenna up to 5 percent either side of the center frequency without any problems, and sometimes 6 or 7 percent.

With a good receiver, a low-noise QTH (location), and a high-power transmitter, a full-size dipole antenna up 30 meters or more in length will allow you to "work the world" over any propagation path that lies on the dark side of the globe. During the daytime, however, no antenna will likely get you contacts over distances greater than 800 kilometers or so.

Allocations by Mode

Most of the HF ham bands have segments designated for use with specific modes. Perhaps better stated, the suballocations restrict usage to certain modes. On this particular band, you may use only digital modes on the lowermost 100 kHz (3.500 MHz to 3.600 MHz). Those modes include CW, RTTY, PSK, MFSK, and data transmission. Other, more exotic digital modes are also allowed.

Tip

On all the ham bands, CW emission is allowed from top to bottom. That is to say, you can send and receive Morse code legally on any amateur frequency! However, for practical reasons, nearly all CW enthusiasts stay within the digital segments of bands that are divided up by mode; and even then, they prefer the lowermost frequencies within those subbands, leaving the upper portions to those who prefer RTTY, PSK, MFSK, and the like.

On the "lion's share" of the 80-and-75-meter band, going from 3.600 MHz to 4.000 MHz, voice (also called *phone*) and image emissions are allowed. So-called "75-meter sideband" or "75-meter phone" enjoys huge popularity in the United States, especially on winter nights when you can hear signals from all over the continent with ease. Often the band gets downright crowded! Always remember to stay within the subbands allowed for your license level.

Allocations by License Class

The 80-and-75-meter band contains various segments allocated according to license class. Extra Class operators can use all the bands from top to bottom, including this one. Other license classes have privileges as follows:

- Advanced Class hams can use the ranges from 3.525 MHz to 3.600 MHz for digital operation, and 3.700 MHz to 4.000 MHz for phone and imaging communications.
- General Class hams can use the ranges from 3.525 MHz to 3.600 MHz for digital operation, and 3.800 MHz to 4.000 MHz for phone and imaging.
- Novices, Technicians, and Tech Plus operators can use the frequencies from 3.525 MHz to 3.600 MHz for CW operation only, with a maximum of 200 watts *peak envelope power* (PEP) output.

What's PEP?

If you haven't heard or read about PEP, here's a quick definition: It's the maximum instantaneous RF signal power that your transmitter puts out. In technical terms, it's the instantaneous power at the peaks of the modulation envelope.

The ratio of PEP to the *average power* depends on the emission mode. If you send out a constant, unmodulated, pure carrier, the PEP equals the average power. The same holds true for FSK and PSK modes. For CW, it varies a little bit, depending on your sending quirks, but usually the average power is about 40 to 50 percent of the PEP. For SSB voice without compression or clipping or other processing, the ratio hovers around 25 to 33 percent. For FM, as with FSK and PSK, it's 100 percent (1 to 1).

Heads Up!

No ham may use a transmitter that puts out more than 1500 watts (or 1.500 kilowatts) PEP on any band or mode, ever. On some bands and for some license classes, further restrictions apply. These constraints keep changing, so for the latest information, I recommend that you visit the ARRL website at

www.arrl.org

The FCC rules also state that you must use the lowest amount of output power necessary to carry on the communications that you want to do. So in fact, even if the legal limit is 1500 watts but you can get away with 150 watts to maintain a contact, then technically, you should scale back your "rig" to 150 watts output.

Table 4-2 breaks down the 80-and-75-meter band according to license class suballocations and mode usage restrictions.

TABLE 4-2 Suballocations in the United States for the 80-and-75-meter band (3.500 MHz to 4.000 MHz) as of March 2012. Please note that this information will likely change in the coming years, so you might want to double-check this table against the ARRL website or the latest edition of *The ARRL Handbook*.

Frequency Range	Modes Allowed	License Level Restrictions
3.500 MHz to 3.525 MHz	CW and digital	Extra
3.525 MHz to 3.600 MHz	CW and digital	Extra, Advanced, and General
3.525 MHz to 3.600 MHz	CW	Novice, Technician, and Technician Plus limited to 200 watts; Extra, Advanced, and General up to 1500 watts
3.600 MHz to 3.700 MHz	CW, non-FM voice, and imaging	Extra
3.700 MHz to 3.800 MHz	CW, non-FM voice, and imaging	Extra and Advanced
3.800 MHz to 4.000 MHz	CW, non-FM voice, and imaging	Extra, Advanced, and General

60 Meters

The band known as 60 meters does not comprise a continuous span of frequencies as the other ham bands do; instead, it has well-defined channels, each one 3.000 kHz wide and centered at the following frequencies as of this writing:

- 5.3320 MHz
- 5.3480 MHz
- 5.3585 MHz
- 5.3730 MHz
- 5.4050 MHz

Your signal can't legally exceed 2.800 kHz in bandwidth, and you must center your signal to coincide precisely with the channel center. This notion is simple for digital modes but a little more complicated for USB. You don't want to set your transmitter's *suppressed carrier* frequency to coincide with a channel center. If you do that, part of your emitted energy will stray outside the channel on the high end. Ideally, you should "tweak" your suppressed carrier frequency so that it lies 1.500 kHz *below* the channel center. Then, if your radio is typical and has an audio passband of 300 Hz to 3000 Hz and an RF filter that trims your signal to 2.700 kHz of bandwidth, your energy will lie entirely within the channel; in fact, you'll have a little bit of room to spare on the low end.

For Techies Only

If you have access to a *spectrum monitor,* such as the sort you might find in a high-end HF communications receiver or a professional electronics lab, you can use it to make sure that you stay inside the allocated channel.

Sharing with Other Services

Hams must share the 60-meter band with other services, taking pains to avoid interfering with those services. Only one signal may exist at a given time, audible from a given location, on a given channel. So if you hear someone having a contact on a particular channel, don't transmit there.

Characteristics

Like 80 meters and 160 meters, the 60-meter band works better for long-haul contacts when most, or all, of the signal travels over the dark side of the earth. However, the effect is not quite as pronounced. Once in a while, you'll hear stations from 1600 kilometers to 3200 kilometers away during the daytime, especially if the sun rests low in the sky.

Also like 80 meters and 160 meters, the part of the spectrum around 5 MHz suffers adverse effects on account of sferics during the thunderstorm season. However, the problem is not quite so severe or extensive here as it is on those lower bands. In general, as the frequency goes up, sferics propagate for shorter and shorter distances before they fade down enough to give you a break! So don't let the fact that it's a summer day in the American Midwest deter you from listening on this band.

You can get an idea of the propagation conditions between Colorado, USA and your QTH on 60 meters at any time by tuning your radio to 5.000 MHz and listening for the time broadcast station WWV, operated by the *National Institute of Standards and Technology* (NIST). You can also find out about propagation conditions between Hawaii and your location by looking out for WWVH on the same frequency. Once in a while, you'll hear both stations at the same time, competing with each other!

Allocations by Mode

You may use USB mode for voice (not LSB), as well as the common digital modes, such as CW, RTTY, PSK, on 60 meters. Old-fashioned AM voice and FM voice are forbidden. Your transmitter must not put out more than 100 watts PEP.

Allocations by License Class

Anyone who holds an Extra, Advanced, or General class license may use any channel in the 60-meter band, keeping in mind the restrictions outlined above.

40 Meters

The 40-meter band spans 7.000 MHz to 7.300 MHz. In free space, a 40-meter-long wave has a frequency of 7.500 MHz, slightly above the ham band, whose center frequency actually lies just under 42 meters.

Sharing with Other Services

Amateur Radio operators enjoy full privileges on this band, not legally having to yield to anybody else. But the theory of law and practical reality sometimes diverge, and this band offers a good example of that circumstance. The upper 1/3 of the band, at frequencies between 7.200 MHz and 7.300 MHz, harbors high-powered AM shortwave broadcast stations scattered around the globe. Hams using SSB in this part of the band will notice these signals as they pass through receiver product detectors; they'll sound like heterodynes or "beat notes" with space-alien-like "monkey chatter" superimposed.

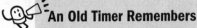

An Old Timer Remembers

The shortwave broadcast stations can get annoying in the upper 100 kHz of 40 meters. But if you're a new ham, consider yourself lucky. Ignorance is bliss: In the 1960s and 1970s, the problem was far worse. Not only did more of the 40-meter band serve as home to shortwave broadcast stations, but also it seems (if my memory serves) that there were more of them per unit spectrum span. On some nights, I could hardly use the band because of interference from these stations. On more than one occasion, I switched my radio to the band, listened for about two minutes, gave up in favor of either 80 meters or 20 meters, or else left the ham shack and did something else.

Characteristics

The 40-meter ham band treats signals just about the same way as the 41-meter shortwave broadcast band does. During the day, you can communicate reliably at distances of up to about 1600 kilometers. Conditions tend to be a little better in the fall and winter because sferics, and atmospheric noise in general, are less intense than they are in the spring and summer. The band "opens up" at night all year round, except when a solar disturbance disrupts shortwave propagation in general.

On a winter night when solar activity is high but not out of control, the 40-meter band can provide you with hours of fun, even if you can manage only a low dipole as an antenna. A half-wave dipole measures about 20 meters from end to end on this band. If you can get it up at least 10 meters, you'll get fine results. Of course, as with most ham antennas, you should string your antenna up as high as you can.

Tip

Has anyone told you that you can't work DX (far-away foreign stations) on 40 meters without a high-powered linear amplifier? Don't believe that nonsense! I've worked DX on 40-meter CW with 10 watts output and an end-fed wire only a few meters above the ground. So no matter how modest your station is, try this band!

Allocations by Mode

On 40 meters, you may use only digital modes on the lowermost 125 kHz (7.000 MHz to 7.125 MHz), including CW, RTTY, PSK, MFSK, data transmission, and other more exotic digital modes. Between 7.125 MHz and 7.300 MHz, phone and image emissions are allowed. Like its counterpart on 75 meters, "40-meter phone" gets a lot of use. Remember to stay within the subbands allowed for your license class.

Allocations by License Class

Extra Class operators can use the whole 40-meter band, subject to mode restrictions. Other license classes have privileges as follows:

- Advanced Class licensees can use the ranges from 7.025 MHz to 7.125 MHz for digital operation, and 7.125 MHz to 7.300 MHz for phone and imaging communications.
- General Class operators can use the ranges from 7.025 MHz to 7.125 MHz for digital operation, and 7.175 MHz to 7.300 MHz for phone and imaging.
- Novices, Technicians, and Tech Plus operators can use the frequencies from 7.025 MHz to 7.125 MHz for CW operation only, with a maximum of 200 watts output.

An Old Timer Remembers

When I first got my General Class license in March of 1967, the 40-meter phone band went only from 7.200 MHz to 7.300 MHz. The lower 2/3 of the band served as home to CW and RTTY operators only. The narrowband modes got more spectrum space than the wideband modes got! That state of affairs seems bizarre by today's standards, doesn't it? Most of the other HF bands were allocated according to the same "CW-centric" philosophy.

Table 4-3 breaks down the 40-meter band according to license class suballocations and mode usage restrictions.

TABLE 4-3 Suballocations in the United States for the 40-meter band (7.000 MHz to 7.300 MHz) as of March 2012. Please note that this information will likely change in the coming years, so you might want to double-check this table against the ARRL website or the latest edition of *The ARRL Handbook*.

Frequency Range	Modes Allowed	License Level Restrictions
7.000 MHz to 7.025 MHz	CW and digital	Extra
7.025 MHz to 7.125 MHz	CW and digital	Extra, Advanced, and General
7.025 MHz to 7.125 MHz	CW only	Novice, Technician, and Technician Plus limited to 200 watts
7.125 MHz to 7.300 MHz	CW, non-FM voice, and imaging	Extra and Advanced
7.175 MHz to 7.300 MHz	CW, non-FM voice, and imaging	Extra, Advanced, and General

30 Meters

The ham band at 30 meters extends from 10.100 MHz to 10.150 MHz, so it's a relatively small band compared with most other HF bands. Nevertheless, you can have a lot of fun in this little span of frequencies. I remember the days when hams dreamed of having a band here; 40 meters and 20 meters seemed separated by a gigantic gulf. No longer!

Sharing with Other Services

Amateur Radio operators must share this band with fixed services outside the United States. When you operate on 30 meters, you must make sure that you don't interfere with those services.

Characteristics

This band, like the 31-meter shortwave broadcast band, lies in a "transition zone" as you go up in frequency, where the propagation begins to get better in the daytime than at night. You can expect to make contacts all over the world when most, or all, of the path lies in darkness. When most, or all, of the path lies in daylight, you can hear, and be heard, over distances up to around 5000 kilometers on a regular basis. Sferics will bother you less on this band than they do on 80-and-75, 60, or 40 meters.

You can get away with a modest-size antenna on 30 meters. A half-wave dipole spans 14 meters from end to end; if you cut it for 10.125 MHz at the exact center of the band, you'll get a good impedance match to 50-ohm or 75-ohm coaxial cable over the entire range. If you can get that dipole up at least 7 meters, you can expect decent results. A full-size quarter-wave vertical antenna measures about 7 meters tall.

You have an asset here similar to the one you can enjoy on 60 meters. To check the propagation conditions between Colorado, USA and your location on 30 meters, tune to 10.000 MHz and listen for WWV. You might also hear WWVH at the same time as, or instead of, WWV. Both stations identify themselves periodically, so if you have any doubt about which of them you're hearing, listen for a while.

Allocations by Mode

No matter where you operate in the 30-meter band, you must restrict your work to digital modes only, including RTTY, PSK, MFSK, data, and of course, CW. Your transmitter cannot put out more than 200 watts PEP.

Allocations by License Class

If you have a General, Advanced, or Extra Class license, you can operate on any frequency in the band. If you're a Novice, Tech, or Tech Plus operator, you'll have to upgrade at least to General before you can transmit on 30 meters.

20 Meters

The 20-meter band goes from 14.000 MHz to 14.350 MHz. Hams don't have to share this band with any other services. Only General, Advanced, and Extra Class operators have access to this choice span of frequencies, but they can use up to the full legal limit of transmitter output power (1500 watts PEP) throughout.

As one of the best all-around long-range communication bands, 20 meters probably constitutes the single most popular span of frequencies allocated to radio hams today. Because it can get crowded, this band occasionally harbors people who get impatient, and even rude, with their fellow operators.

Once in a while, you'll hear a "lid" (poor operator) on this band who simply acts hostile or irrational. Never reciprocate in kind; maintain your courtesy no matter how rotten somebody else gets.

Be Cool!

On a busy band, such as 20 meters on a good day, always ask if the frequency is in use before you send a CQ (a request for anyone who hears you to answer). On SSB, say "Is the frequency in use? This is (your call letters)." On CW, send "QRL?" followed by a pause and finally "de" and your call sign.

Characteristics

The 20-meter Amateur Radio band behaves rather like the shortwave broadcast 22-meter and 19-meter bands. During the day, you can communicate reliably all over the globe when conditions are decent. On this band and all the HF bands at higher frequencies (shorter wavelengths), the light/dark propagation dichotomy reverses from its lower-frequency state. In other words, "20 meters and down" do better, generally, over sunlit paths than they do over dark paths. However, at and near sunspot cycle peaks, the 20-meter band offers worldwide communications with low power and modest antennas 24 hours a day, especially in the summer at latitudes close to the geographic poles.

On a spring or summer afternoon when solar activity is high but not stormy, the 20-meter band can provide holders of General, Advanced, and Extra Class licenses a good deal of fun, including plenty of DX (long distance to foreign countries). A half-wave dipole measures about 10 meters from end to end on this band. If you can get it up at least 5 meters, it'll work okay. Another alternative is the ground-plane antenna, essentially a quarter-wave vertical with three or four quarter-wave radials and with its base elevated at least a quarter wavelength above the earth's surface. This type of antenna measures roughly 5 meters tall, and the radials measure the same length.

> **Tip**
>
> A quarter-wave vertical "in the clear" and with a good ground system can provide better DX performance, in some cases, than a dipole can do. That's because a well-engineered vertical offers a low angle of radiation and reception, a virtual necessity for DX operation.

You can get a good clue as to the conditions between Colorado or Hawaii and your location on 20 meters by setting your radio to 15.000 MHz and listening for WWV and/or WWVH. If you hear only one of them, make sure that you know which one is coming in! You might get a surprise when you think you're hearing WWV and, in fact, it's WWVH (or vice-versa).

Allocations by Mode

You may use only digital modes on the lowermost 150 kHz (14.000 MHz to 14.150 MHz), including CW, RTTY, PSK, MFSK, data transmission, and their "cousins." Between 14.150 MHz and 14.350 MHz, you may put out phone and image emissions. Of course, you must adhere to the license class restrictions when choosing a frequency.

Allocations by License Class

Extra Class operators can use the whole 20-meter band, subject to mode restrictions. Other license classes have privileges as follows:

- Advanced Class licensees can use the ranges from 14.025 MHz to 14.150 MHz for digital operation, and 14.175 MHz to 14.350 MHz for phone and imaging operation.
- General Class operators can use the spans from 14.025 MHz to 14.150 MHz for digital operation, and 14.225 MHz to 14.350 MHz for phone and imaging contacts.
- The 20-meter band is off-limits to Novices, Technicians, and Tech Plus operators. The prospect of getting on this band has motivated a lot of people to upgrade!

Table 4-4 breaks down the 20-meter band according to license class suballocations and mode usage restrictions.

17 Meters

Ham radio operators can use the frequency span from 18.068 to 18.168 MHz, commonly called 17 meters. For experimenters and technophiles, it's a neat slot between the larger, more crowded, and frequently contest-cluttered 20-meter and

TABLE 4-4   Suballocations in the United States for the 20-meter band (14.000 MHz to 14.350 MHz) as of March 2012. Please note that this information will likely change in the coming years, so you might want to double-check this table against the ARRL website or the latest edition of *The ARRL Handbook*.

Frequency Range	Modes Allowed	License Level Restrictions
14.000 MHz to 14.025 MHz	CW and digital	Extra
14.025 MHz to 14.150 MHz	CW and digital	Extra, Advanced, and General
14.150 MHz to 14.175 MHz	CW, non-FM voice, and imaging	Extra
14.175 MHz to 14.225 MHz	CW, non-FM voice, and imaging	Extra and Advanced
14.225 MHz to 14.350 MHz	CW, non-FM voice, and imaging	Extra, Advanced, and General

15-meter bands. It's great for DX, too! Hams don't have to share 17 meters as they do with 30 meters, and full legal power is allowed.

Characteristics

This band behaves like the 16-meter and 15-meter shortwave broadcast bands, which lie close to it in frequency. You can make contacts all over the world when most, or all, of the path lies on the daylight side of the planet and conditions are decent. Sferics will give you far less grief here than on lower bands. In fact, unless a thunderstorm looms on your doorstep, you won't hear much, if any, atmospheric "static" on 17 meters. Humanmade noise presents a different conundrum. I, for one, have trouble with rogue neighborhood electrical appliances on this band.

You can get away with a rather small antenna on 17 meters. A half-wave dipole spans only about 8 meters from end to end; if you cut it for 18.118 MHz at the band center, you'll get a good match to 50-ohm or 75-ohm coax over the whole band. If you can get that dipole up at least 4 meters, you can expect fair results, but if you want to work DX, you should elevate it at least 8 meters above the ground. A quarter-wave vertical measures only about 4 meters tall on this band.

A simple ground-plane antenna is an excellent option on 17 meters for those with a serious DX interest and a modest budget. The radiating element will be about 4 meters tall, and the radials, which can slope down at an angle and serve as guy wires, will also measure roughly 4 meters long, kept at the proper length with egg insulators. If you put the feed point above the ground at least 4 meters, the whole structure will measure only 8 meters from the mounting surface to the tip of the radiator!

> **Tip**
>
> By trial and error, trim the radiating element and the radials of your ground-plane antenna to get optimum performance throughout this band. Make the elements a little too long as you assemble the antenna, and then cut the elements back, centimeter by centimeter, using a diagonal cutter, until your standing-wave-ratio (SWR) meter indicates minimum reflected power at 18.118 MHz.

Allocations by Mode

On the 17-meter band, you must restrict your work to digital modes only, including RTTY, PSK, MFSK, data, and CW, below 18.110 MHz. Above that frequency, you can use phone and imaging emissions.

Allocations by License Class

If you have a General, Advanced, or Extra Class license, you can operate on any frequency within the 17-meter band. If you're a Novice, Tech, or Tech Plus operator, you'll have to upgrade at least to General before you can transmit here.

15 Meters

The 15-meter Amateur Radio band occupies the spectrum span from 21.000 MHz to 21.450 MHz. It's almost entirely a daylight-path band; at night the upper layers of the ionosphere rarely have enough density to return signals to the surface, although exceptions do occur, especially near sunspot cycle peaks. In terms of absolute spectrum space, 15 meters is almost as large as the 80-and-75-meter band. Hams have full privileges here, without any need to worry about interfering with other services.

Characteristics

The ham 15-meter band behaves like the shortwave broadcast 13-meter band. It's a great place in the spectrum for working DX when conditions allow. You don't need much transmitter output power to "work the world" even with a modest antenna, such as a dipole or ground plane, and those antennas have manageable dimensions. A half-wave dipole measures only about 7 meters from end to end, and a full-size, quarter-wave vertical rises only about 3.5 meters up from the feed point.

On 15 meters, complex directional antennas become mechanically manageable even for the ham without civil engineering experience, providing gain (extra transmitted power) in favored directions and attenuation (suppression) of received signals from unwanted directions. Lots of hams use Yagi antennas, also called *beams*, on this band. In addition to having reasonable size themselves, antennas on 15 meters don't have to be mounted very far above the surface to provide excellent

DX performance. If you can get your antenna 15 meters above the ground, it'll sit a full wavelength up there, and you'll get excellent low-angle radiation and reception.

Tip

In Chap. 8, you'll learn about the antennas that ham radio operators favor. If you don't know exactly what a Yagi is and you can't wait to find out, flip ahead in the book and glance at them now, and then dig into the details later.

Allocations by Mode

Amateur Radio operators may legally use only digital and data modes between 21.000 MHz and 21.200 MHz, including CW, RTTY, PSK, MFSK, data, and similar emissions. Between 21.200 MHz and 21.450 MHz, hams are allowed to transmit with phone and image emissions, subject to license class restrictions.

Allocations by License Class

Extra Class operators can use the whole 20-meter band, subject to mode constraints. Other license classes have privileges as follows:

- Advanced Class licensees can use the ranges from 21.025 MHz to 21.200 MHz for digital operation, and 21.225 MHz to 21.450 MHz for phone and imaging communications.
- General Class operators can use the spans from 21.025 MHz to 21.200 MHz for digital operation, and 21.275 MHz to 21.450 MHz for phone and imaging contacts.
- Novices, Technicians, and Tech Plus operators can use the frequencies from 21.025 MHz to 21.200 MHz for CW operation only, limited to 200 watts RF output power.

Table 4-5 breaks down the 15-meter band according to license class suballocations and mode usage restrictions.

An Old Timer Remembers

A 40-meter dipole antenna fed with 50-ohm or 75-ohm coaxial cable will work on 15 meters without modification. At 21 MHz, a 7-MHz dipole forms a one-and-a-half-wave center-fed antenna. So if you have a 40-meter dipole strung up right now, try it on 15 meters and see what happens. The first time I tried that trick as a Novice class operator in 1966 after getting a tip from my mentor Bill, W0GLE, I sent a CQ and snagged some instant DX! Well, okay, VE3 (Ontario), from my station in Minnesota— but at the time, it thrilled me half to death.

TABLE 4-5   Suballocations in the United States for the 15-meter band (21.000 MHz to 21.450 MHz) as of March 2012. Please note that this information will likely change in the coming years, so you might want to double-check this table against the ARRL website or the latest edition of *The ARRL Handbook*.

Frequency Range	Modes Allowed	License Level Restrictions
21.000 MHz to 21.025 MHz	CW and digital	Extra
21.025 MHz to 21.200 MHz	CW and digital	Extra, Advanced, and General
21.025 MHz to 21.200 MHz	CW Only	Novice, Technician, and Technician Plus limited to 200 watts; Extra, Advanced, and General up to 1500 watts
21.200 MHz to 21.225 MHz	CW, non-FM voice, and imaging	Extra
21.225 MHz to 21.275 MHz	CW, non-FM voice, and imaging	Extra and Advanced
21.275 MHz to 21.450 MHz	CW, non-FM voice, and imaging	Extra, Advanced, and General

12 Meters

The 12-meter Amateur Radio band covers 24.890 MHz to 24.990 MHz. This band is actually closer to 25 MHz than to 24 MHz, and lies about midway between 15 meters and 10 meters. It's a good place for experimenters and serious DXers. The full legal limit is allowed, and there's no sharing, either!

Characteristics

This band behaves like the 11-meter shortwave broadcast band. You can make contacts all over the world near the times of sunspot maxima when most, or all, of the path lies on the sunlit side of the earth. Sferics almost never pose a problem; if you hear "thunderstorm static" on 12 meters, you'd do well to go outside and see if a dark cloud is about to come over you.

Tip

Humanmade noise from miscreant home appliances knows no limit on any ham band, and 12 meters is no exception. If you suffer from this sort of problem, as I do from time to time, consider a radio-frequency interference (RFI) solution. Often the problem must be resolved in, or at the location of, the offending device. A company called *Palomar Engineers* offers a set of products that might help you in this kind of situation. Go to

http://palomar-engineers.com

and hit the link "RFI Kits" (or something along those lines). Internet sites and content change on an almost daily basis, but I found this link valid and noted it in December of 2013.

A half-wave dipole for 12 meters spans roughly 5.7 meters from end to end; if you cut it for 24.940 MHz at the band center, you'll get a good match to 50-ohm or 75-ohm cable over the whole band. You can make lots of contacts if you can get that dipole outdoors and high enough so that a tall person won't run into it.

A ground-plane antenna can yield amazing DX results on 12 meters when propagation conditions are good, and you can build one with hardware-store parts! The radiating element measures about 2.9 meters tall. You can cut radials that double as guy wires to the same length as the radiating element. You'll have to "tweak" the element lengths a little to get the lowest SWR possible at 24.940 MHz.

Tip

With a ground-plane antenna, you can cut the radials to 1/4 wavelength and then leave them alone, trimming only the radiating element. To calculate 1/4 electrical wavelength in meters for tubing or wire, divide 71.3 by the frequency in megahertz. Use telescoping lengths of tubing for the radiating element. Lengthen and shorten it centimeter by centimeter, sliding one of the sections out of or into its neighbor, until you get minimum SWR at the desired frequency. If the radials end up slightly longer or shorter than the radiating element when you've completed the adjustment process, the discrepancy won't affect the performance of the antenna in real terms, so don't worry about it. Also, if your minimum SWR isn't a perfect 1:1 match, don't fret about that either. As long as it's 2:1 or less, you should not have any trouble.

Allocations by Mode

On 12 meters, you must restrict your work to digital modes only, including RTTY, PSK, MFSK, data, and CW, between 24.890 MHz and 24.930 MHz. Above 24.930 MHz and all the way up to 24.990 MHz, you can use phone and imaging emissions.

Allocations by License Class

If you have a General, Advanced, or Extra Class license, you can operate over the whole 12-meter band. If you're a Novice, Tech, or Tech Plus operator, you'll have to upgrade at least to General before you can fire up your transmitter here, except into a *dummy load*.

What's a Dummy Load?

When you want to test a transmitter at full RF output but you don't want your signal to go over the air, you can connect a dummy load to its antenna output terminals. This contraption comprises a noninductive resistor with a power dissipation rating high enough to convert all of your transmitter's RF energy to heat without burning up. Use your favorite search engine with the phrase "dummy load" if you want to shop for one! I recommend that you use a dummy load whenever possible, so as to avoid causing interference to your fellow hams on the air. A good dummy load is also a great addition to the experimenter's repertoire!

10 Meters

If you've had experience with Citizens Band (CB) radios, you should have a good idea of how the 10-meter ham band treats signals. It's a big band, extending from 28.000 MHz to 29.700 MHz. Amateur Radio operators enjoy unrestricted use of the band; it involves no sharing with other services. If you have a General, Advanced, or Extra Class license, you can use any frequency in the band and put out up to 1500 watts PEP.

Characteristics

Because of its ability to provide spectacular worldwide communications with low transmitter power output under ideal conditions, 10 meters has acquired the nickname "magic band." Over daylight paths, especially in spring and summer and in years near the sunspot cycle maximum, you can work DX with less than a watt of RF at your antenna. And the antenna itself can be modest; a ground plane or dipole only a few meters above the ground will do the job! Of course, you can run higher power and use more sophisticated antennas if you want, but QRP (low power) operation with simple antennas has proven great fun for a lot of technically oriented hams, myself included.

On the downside, the 10-meter band spends a great deal of time in "hibernation," when it behaves more like a VHF band than an HF band. In the wintertime, at night, or during sunspot minima, you will often find that contacts are possible only up to distances of 30 to 40 kilometers on this band. Ionospheric propagation goes away entirely. Nevertheless, you should never assume that 10 meters is dead just because you don't hear any signals there. Sometimes people, hearing nothing on 10 meters, assume it's dead when in fact it's wide open! I call this psychological phenomenon the "Dead Band Delusion." So why not send a CQ or two?

Allocations by Mode

Only the digital modes (RTTY, PSK, MFSK, data, and CW) are allowed between 28.000 MHz and 28.300 MHz. From 28.300 MHz to 29.700 MHz, you can use voice and imaging emissions.

Allocations by License Class

If you have a General, Advanced, or Extra Class license, you can operate over the whole 10-meter band using the emissions as described above. If you're a Novice, Tech, or Tech Plus operator, you can use any of the digital modes between 28.000 and 28.300 MHz, and SSB phone between 28.300 MHz and 28.500 MHz, keeping your power output at or under 200 watts PEP.

An Old Timer Remembers

When I served as an editorial assistant at ARRL Headquarters in 1978–1979, I had an apartment in West Hartford, Connecticut, on the top floor of an old three-story house. My landlord had a good attitude when I asked him if I could put an antenna on the roof. A colleague and I got hold of a 3-element Citizens Band (CB) Yagi antenna kit, and assembled it with all the dimensions reduced by 5 percent to make it work at 28 MHz instead of 27 MHz. We took the instructions and simply multiplied every measurement number by 0.95! A television antenna rotator and a 3-meter-tall mast completed the ensemble. That Yagi had a "clear view" of the horizon in all directions from its vantage point on that sloping roof. When I got on 10-meter CW and dialed my Drake T-4X transmitter down to 3 watts RF output, I snagged a New Zealand contact on CW and got a signal report of S7 (moderately strong). In my mind and memory, the 10-meter band earned its reputation as a magic band on that day!

How Do They Look?

Figure 4-1 shows the ham radio bands below 30 MHz, on a linear scale according to frequency. The frequency goes up, and the wavelength gets shorter as you move up the scale. Although propagation varies from band to band, conditions at the top of any given band are nearly always the same as conditions at the bottom of the same band. Notice that while ham radio operators enjoy operating privileges in numerous bands that cover an excellent overall sampling of the HF spectrum, the Amateur Radio "slices," in terms of sheer quantity, add up to only a small part of the "whole pie."

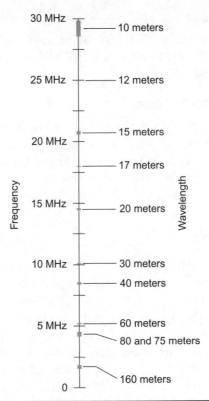

FIGURE 4-1   The ham radio bands below 30 MHz as of 2013, on a linear scale according to frequency.

6 Meters

Amateur Radio operators have 6-meter privileges between 50.000 MHz and 54.000 MHz. In terms of frequency, 6 meters constitutes the lowest of the three VHF ham bands, the others being 2 meters and 1.25 meters. Like 10 meters, this band can seem magical, opening up to the world with ionospheric propagation every once in a while. Such events are rare, but if you listen on this band long enough, you'll see some cool things happen here. Radio hams have the whole band to themselves, and can use up to the legal power limit.

Characteristics

Most of the time, 6 meters offers line-of-sight communications up to 30 or 40 kilometers on a reliable basis. Tropospheric scattering and bending often allow for contacts over greater distances, sometimes several hundred kilometers. Weather fronts can enhance the bending effect and also allow for occasional contacts as a result of ducting, in which radio waves get trapped within a layer of cold air sandwiched between two layers of warmer air. When ducting takes place on a grand scale, contacts can sometimes happen over distances of up to about 1500 kilometers.

Have You Heard?

The commercial FM broadcast band at 88 MHz to 108 MHz in the United States can be affected by ducting in a dramatic way every now and then. Have you ever heard a broadcast station suddenly come in from, say, Kansas City as you ran around Lake Harriet in Minneapolis with a portable FM headset radio? If so, you've witnessed the duct effect!

Ionospheric F layer propagation has occurred on 6 meters. During that sort of opening, you can communicate in much the same way as you can do on 10 meters under similar circumstances. More often, 6-meter ionospheric propagation takes place via the E layer, which tends to ionize in "clouds" at an altitude of around 80 kilometers during periods of sunspot maxima. This effect, called *sporadic-E propagation*, happens quite a lot over *transequatorial paths* (paths that cross the earth's equator) a few weeks either side of the *equinoxes*, when the sun lies in the same plane as the earth's equator. Look for transequatorial sporadic-E openings in March, April, September, and October.

More exotic modes of communication, such as auroral propagation and meteor scatter, occasionally manifest themselves here. These modes generally require CW or a synchronized digital mode.

Allocations by Mode

The lowest 100 kHz of the 6-meter band, from 50.000 MHz to 50.100 MHz, is reserved for CW operation only. Most of the CW operation that I've heard takes place around 50.090 MHz. You'll occasionally hear beacon stations on other CW frequencies. You'll recognize them by their relatively slow, repetitive transmissions with call signs given once a minute or so. The rest of the band, from 50.100 MHz all the way up to 54.000 MHz, is open to all modes of operation that hams commonly use, with the exception of fast-scan television.

Allocations by License Class

All hams except Novices may legally use any frequency in the 6-meter band, subject to the mode constraints outlined above. If you hold an old Novice Class license, you must upgrade at least to Tech before you can use this band.

2 Meters

The slice of spectrum from 144.000 MHz to 148.000 MHz forms the ham radio 2-meter band. It's one of the most (if not the most) popular ham bands, with the majority of operation done with FM transceivers and repeaters. Hams don't have to share this band with any other services.

Characteristics

Under most circumstances, 2-meter communications happens in a line-of-sight mode, resembling the bands at higher frequencies. However, tropo can take place here, in much the same way as it does on 6 meters. Ionospheric F-layer propagation has not been observed. Meteor scatter and auroral communications can be carried out, but they're a little more difficult to initiate and maintain than they are on 6 meters.

Since the "repeater revolution" in the 1970s, propagation forecasts and conditions (and the science that goes along with it) are of little or no concern to radio hams on 2 meters. You might use the band exclusively for repeater communications for years, only to get a rare surprise when a tropo event pops up and you hear someone 800 kilometers away, coming in as if they were located in your neighborhood!

Tip

When a tropo event occurs on 2 meters, the signals can actuate distant repeaters and produce chaos unless those repeaters are closed to all signals except authorized ones. Tone squelching has largely solved this occasional problem, but it can also keep out-of-town motorists from accessing the repeater in an emergency unless they know the requisite tone frequency.

Perhaps the "coolest" mode that hams use on 2 meters—pretty much unknown on the longer-wavelength bands—is earth-moon-earth (moonbounce). To make moonbounce contacts, you'll need a transmitter that can put out a lot of RF power, preferably the legal limit of 1500 watts PEP, and work in CW or one of the synchronized digital modes. You'll also need a high-gain directional antenna and a sensitive receiver. And finally, if you live in an RF-noisy location as I do, you might as well forget about moonbounce! The best locations have underground electrical lines and residential lots big enough so that neighbors with rogue electrical appliances can't ruin reception. You'll want to check your local zoning laws concerning large antennas, and make sure that none of your neighbors will hate you when they see a matrix of 2-meter Yagis sprout in your backyard.

Allocations by Mode

You may use only CW emission from 144.000 MHz to 144.100 MHz. The range from 144.100 MHz to 148.000 MHz is open to all modes of operation that hams commonly employ, but as on the lower bands, fast-scan television is forbidden because it takes up too much bandwidth!

Allocations by License Class

If you hold an Extra, Advanced, General, or Tech license, you may use any frequency in the 2-meter band, subject to the above-described mode restrictions. If you hold a Novice license, you must upgrade before you can transmit on 2 meters. The prospect of communicating with handheld and mobile radios, along with repeaters, has probably been the greatest factor to motivate old Novices to become Techs!

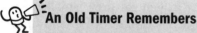

An Old Timer Remembers

During my Novice year in 1966–1967, the span from 145.000 MHz to 147.000 MHz formed a 2-meter allocation for that class. As on the HF Novice bands, I had to use a crystal-controlled transmitter that put out no more than 75 watts. But I could use voice emissions there, and I wanted to try some on-the-air "gum flapping" instead of "brass pounding"! Repeaters hardly existed at all in the mid-1960s, so most operations on 2 meters were done using AM. A friend loaned me a homemade vacuum-tube, 2-meter AM rig that had about the same dimensions and weight as a small refrigerator (and probably gobbled up as much electricity too). I plugged a quartz crystal and a microphone into that "boat anchor," connected my 40-meter dipole to it (not knowing much about antennas yet), and chatted with other hams around town as I sat in the family fallout shelter that served as my "ham shack."

Beyond 2 Meters

At frequencies above 148 MHz, radio hams enjoy privileges on numerous small slices of the spectrum, as shown in Table 4-1. Once you get up to 275 GHz, you can communicate on any frequency you want, including microwaves, infrared (IR), visible light, ultraviolet (UV), X-rays, and even gamma rays, assuming that you can come up with a transmitter and receiver that will work at those microscopic wavelengths!

Did You Know?

Some radio hams, myself included, have built modulated-light transmitters and receivers, communicating over distances ranging from a few meters to a few kilometers. Of course, such communications have to take place in a line-of-sight mode—literally!

Characteristics

Tropo sometimes happens on 1.25 meters and 70 centimeters, and rarely on bands at higher frequencies. Moonbounce has gained considerable popularity on 70 centimeters, and some hams do it on 23 centimeters as well. Satellite links have grown increasingly common in recent years. Repeaters exist on 1.25 meters and especially on 70 centimeters, similar to the ones on 2 meters.

As the wavelength grows shorter, antennas in general grow smaller for a given amount of gain. Large Yagis are practical on 219.000 MHz and above; more exotic antennas, such as helical types, horns, and dishes, appear at 23 centimeters and shorter wavelengths. Hams often combine UHF antennas in arrays called *bays*—for example, four helical antennas at the corners of a square, or even nine of them at the points of a "Tic-Tac-Toe" matrix! When fed in phase and carefully aligned, such bays can have directionality and gain comparable to a dish.

Allocations by Mode

On 1.25 meters, the range from 219.000 MHz to 220.000 MHz is reserved for fixed digital message forwarding systems. Otherwise, hams can use all known modes throughout all bands, except they can't legally use fast-scan television (FSTV) on 1.25 meters. In order to use that mode, you have to stay above 420 MHz. Full legal power is allowed except for Novices in their subbands on 1.25 meters and 70 centimeters.

Allocations by License Class

Extra, Advanced, General, and Tech operators may use any frequency on any allocated ham band above 219.000 MHz, subject to the above-described restriction on FSTV. Novices have slices of 1.25 meters (222.000 MHz to 225.000 MHz) and 23 centimeters (1270.000 MHz to 1295.000 MHz). On 1.25 meters, Novices are limited to 25 watts PEP output, and on 23 centimeters, to 5 watts PEP output. However, they can use any legally authorized Amateur Radio mode in those slots.

Heads Up!

Depending on where you live, you might have to adhere to special limitations on operating frequencies and maximum output power. For details on the latest regulations on these matters, contact the ARRL.

Stay Tuned!

For decades, ham radio operators have longed to regain some of the privileges they had in the earliest days of wireless communications more than 100 years ago. Most notable are the frequencies below the standard AM broadcast band, whose lower boundary lies at 535 kHz. A couple of small LF bands have been proposed recently, and radio hams will probably gain privileges "down there" before too long. When and if hams get any new bands, the ARRL will make them known right away!

CHAPTER 5

Fixed Ham Stations

Imagine that you just got your General Class license and bought some equipment for a station. You have a transceiver, an antenna tuner, a computer, and an interface unit that'll let you use some of those fancy digital modes on the HF bands! You've unpacked the equipment from the boxes. You've obtained all the interconnecting cables that you'll need, and you're ready to put it all together and get on the air.

Where Will You Put Your Rig?

Where in your home do you want your rig to reside? You have several choices from the basement to the top floor.

The Electrical System

Before you decide where to put your new ham radio station, you'll want to make sure that you can get enough electricity to the equipment. If you have a low-power or medium-power transmitter (150 watts output or less) and a notebook computer, you'll do okay with a common utility outlet on a circuit rated at 15 or 20 amps. In the United States, the standard AC utility provides 117 volts, give or take a few percent, so 15 amps would give you the ability to power a rig that demands up to 117 volts × 15 amps = 1755 volt-amps. That's around 1600 or 1700 watts, assuming that you don't connect anything else to the circuit. You can find outlets like that just about anywhere these days.

If you want to run high power and use an outboard RF power amplifier (called a *linear amplifier,* or simply a *linear*), you'll want to have a 234-volt circuit for your rig, rather than a 117-volt circuit. The higher voltage circuits take advantage of both components of the split-phase electrical system that most homes have in the United States. You'll get twice the power for the same number of amps, and you'll get a more stable voltage as well.

> **Tip**
>
> If you "load down" a 117-volt circuit so that it must provide its full rated current (or close to it), you'll notice that the voltage drops when you key up a linear. It's the same effect that you get if you run an electric space heater on the high-heat setting.

Whether you go for a 117-volt system or a 234-volt system, make sure that the power line has a good electrical ground connection. That means you'll need outlets with three slots as well as power cords with three conductors and three prongs in their plugs. If you have a 117-volt installation, you can add a power strip with a *transient suppressor* (sometimes mistakenly called a "surge protector") and a 15-amp breaker. Then, once you have met all those requirements, you must make sure that the ground slot in your outlet actually goes to the household ground.

People often assume that a three-wire 117-volt utility outlet has a good ground at its "third hole" (the bottom hole, not either of the vertical slots). That's not always true. I've seen outlets in which that "third hole" wasn't connected to anything! You can use a long extension cord and a *volt-ohm-milliammeter* (VOM), also called a *multimeter*, to find out whether or not a particular three-wire 117-volt outlet has a good electrical ground at its "third hole." Go through the following steps using Fig. 5-1 as a reference. If you want this test to work, you'll have to find a reference outlet in your house that you know has a good ground at its "third hole."

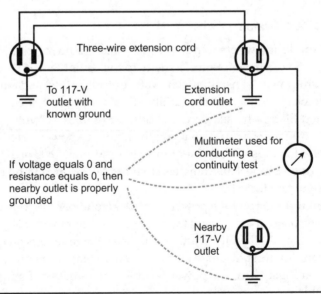

FIGURE 5-1   Arrangement for determining whether the "third prong" of an outlet actually goes to electrical ground. Follow the procedure for a continuity test. Always check for voltage before measuring resistance! Wear your gloves at all times.

1. Put on a pair of rubber gloves and a pair of shoes with soles that will electrically insulate your feet from the floor.
2. Plug a long three-wire extension cord into an outlet with a known electrical ground at the "third hole."
3. Locate the outlet end of the extension cord next to the outlet whose "third hole" you want to test.
4. Set your multimeter to measure the highest AC voltage that it can deal with.
5. Insert the multimeter's black (negative) probe tip into the "third hole" in the extension cord outlet.
6. Insert the multimeter's red (positive) probe tip into the "third hole" of the outlet you want to test, while leaving the black probe tip in the extension cord outlet.
7. Check the meter reading. If it shows anything other than 0, then an AC voltage exists between the two points, so your outlet *does not* have a good ground. It will present a shock or fire hazard if you use it.
8. If you see 0 as the result for step 7, switch the meter to the next lower AC voltage function and repeat the test.
9. If you get 0 again, repeat steps 8 and 9 with all the AC voltage functions that the meter has, going down until you've tested at the lowest AC voltage setting. You should always get a reading of 0. If you don't, then you know that some AC voltage exists between the two points, so your outlet *does not* have a good ground.
10. Assuming that you've seen readings of 0 for all the AC voltage settings, repeat the tests with your multimeter's DC voltage settings, starting with the highest one and working your way down to the lowest one. Test for DC voltage in both directions: first with the black probe tip in the "third hole" of the cord outlet and the red probe tip in the "third hole" of the outlet under test, and then the other way around. You should always see a meter reading of 0.
11. If you ever see any DC voltage besides 0 between the two points, then you know that your outlet *does not* have a good ground.
12. If you see 0 for all of the AC and DC voltage results, switch your multimeter to the highest resistance function.
13. Touch the meter's two test probe tips to each other. If you're using an analog meter, tweak the "0Ω ADJ" knob so that the meter indicates 0. If you're using a digital meter, make sure that the display indicates a value of 0.
14. Insert the black probe tip back into the extension cord's "third hole," and the red probe tip back into the outlet's "third hole." You should get a reading of 0.
15. Reverse the test leads. You should again get a reading of 0.
16. Repeat steps 13 through 15 for all the rest of the meter's resistance settings, working your way down one setting at a time, until you get to the lowest one.

17. If you have observed readings of 0 for every step in the foregoing process, then you can have confidence that your outlet's "third hole" is properly grounded.

18. If you ever see any meter reading other than 0 for the condition between the two "third holes," you know that your outlet *does not* have a good ground. In that case, treat it as a two-wire outlet until you can get a professional electrician to wire it up properly.

Warning!   If the "third hole" in an outlet connects to a wire that isn't grounded, that wire can pick up stray AC by *electromagnetic induction* from other wires in your house, giving you a potentially lethal shock. It can also cause sensitive electronic equipment to malfunction. It can increase the likelihood of, or worsen, any preexisting problems with "RF in the shack." In addition to all that, if you connect a transient suppressor into the circuit, it won't protect anything because it will have no ground to "work against."

Always strive to provide your radios with a steady line voltage. You can tell if the voltage fluctuates under load by connecting an old-fashioned incandescent lamp to the same circuit as your rig. Does the lamp dim when you key up? If so, you need a more robust electrical supply! If you want to get more technical, you can actually hook up an AC voltmeter to your system and watch its reading as you key the rig on and off.

Your choice of rig location will be limited if you want to "run the max" and operate in so-called QRO (high power) mode. You might even need to have an electrician come to your home and install a 234-volt system at the place where you plan to build your station. If you plan to use a minimum amount of power for your station, you won't need to worry about all that electrical business.

> **Tip**
> No matter how low the amount of electrical "juice" that your rig demands from the utility, you should try to avoid powering your equipment through an extension cord. If you must use an extension cord, go to a good hardware store and get one with the largest possible diameter conductors, rated to carry current that equals or exceeds the breaker rating. And, of course, you should keep the cord as short as possible and make sure that it has three conductors.

Basement

Whenever I've lived in a full-sized house, I've located my ham radio equipment in the basement. It seems like the natural place for a hobby like this. There's something about the ambience of a typical basement that lends itself to nerdy activities like ham radio, especially the technical and experimental variety in which I engage. Of course, if you live in a condo or an apartment, or if you live in a part of the country where basements don't work (South Florida, for example), you don't have this option. But it's the one I recommend if you can manage it.

If you opt for the basement, you'll have to figure out how to get your antenna feed line or lines into the radio room (which hams call the *shack*, even if it's an elegant dedicated bedroom or den). That's more difficult for below-ground stations, obviously, than for above-ground stations. When I was first licensed at age 12, my rig went in the family fallout shelter, which my parents had built after the Cuban missile crisis. Mom and Dad didn't want me to drill holes through the concrete blocks for antenna feed lines. So I had to run the feed line around the winding hallway that led into the place, and out into the furnace room, and up through the outside wall of the house just above ground level. That route necessitated a long feed line—over 150 feet of coaxial cable to a 7-MHz dipole 10 feet off the ground. I'd have done better with an open-wire feed line, but it wouldn't have worked with all those bends and nearby copper pipes.

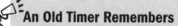 **An Old Timer Remembers**

When I attended college in the 1970s, fears of a global nuclear holocaust had subsided enough so that my parents consented to my drilling two holes through that concrete wall. On some weekends I came "back home" to Rochester from Minneapolis and replaced the coaxial cable with a low-loss open-wire line, one conductor running through each hole, and a shorter and more direct route to a higher antenna.

Tip

You'll read more about feed lines later in this book. Feel free to skip ahead now if you don't know the difference between coaxial line and open-wire line. The gist: Open wire has lower loss than coaxial cable, but it's harder to install properly.

First Floor

The first floor offers several advantages for fixed ham radio stations. You can get a fair RF ground for end-fed, random wires from there by running a wire out a window and connecting it to a ground rod. In addition, the first floors of homes have windows, so you can easily run your feed lines, ground wires, and control cables outside! If you can manage to get your operating desk or table underneath a window, you'll have the best of everything.

The first floor is the most common place for small "dens," which serve well as ham radio zones. If you're married, your spouse might take exception to your demanding an entire room for your radio station, but then again, maybe she (or he) will be happy that you don't want to put that geeky gear out in plain sight, where house guests will see it and cats will make mischief with it. A spare bedroom is another good place for a ham station. Failing all that, you can put it in a corner of the living room.

If your home has an old-fashioned up-and-down sliding window and you can get your radio next to it, you can run feed lines and other wires through the opening using the scheme shown in Fig. 5-2. Get a piece of lumber that has the same thickness as the depression at the base of your window sash. Cut the piece of lumber to a length that equals the inside width of the sash. Then drill holes in the wood, having the correct diameters for your cables and other wires, and thread those cables or wires through the holes (before you install cable or wire connectors, of course). Remove any existing external screen or storm window. Open the window, set the wood with the cables down in the base of the sash, and then close the window, sliding it firmly down to hold the wood in place.

Many hams use transmitters that put out only 100 or 150 watts, so they opt for an end-fed, random wire as the antenna. (If you plan to run the legal limit, an end-fed, random wire will probably give you trouble with "RF in the shack.") When tuned with a good *transmatch* (antenna tuner), an end-fed random wire offers

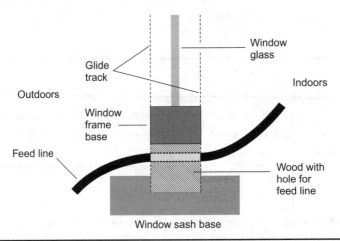

FIGURE 5-2   You can put a strip of wood at the base of an up-and-down sliding window and run feed lines and ground wires through it. You'll have to remove any existing external screen or storm window.

decent performance on any band for which it measures at least an electrical quarter wavelength long. In addition, such an antenna can have a low profile if you use small gauge wire. Unless people know that the antenna exists, they might never even notice it, especially if you can get it up into trees.

Any end-fed wire, no matter how long or short, needs a good RF ground at its feed point. A ground rod by itself won't suffice unless you live in a salt flat or swamp where it rains often! You can enhance the RF performance of your ground system by burying quarter-wavelength radial wires about a foot (30 centimeters) under the surface of your yard. Run them out at various angles. Figure 5-3 shows a hypothetical example for a modest-sized yard. It doesn't matter whether or not the wire has insulation. The ground rod should go down at least 8 feet (2.4 meters), though, and be about 3 feet (1 meter) away from the house foundation.

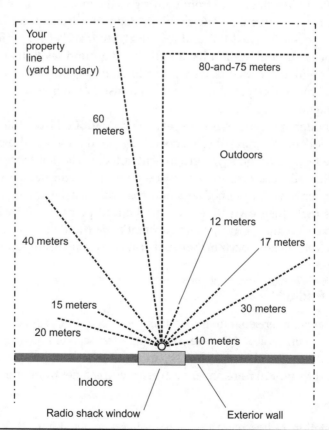

Figure 5-3 Simple method of getting an RF ground for an end-fed, random wire when your station is next to a first-floor window. The small open circle represents the ground rod; heavy dashed lines represent quarter-wavelength buried radials.

> **Tip**
>
> To determine the length of an RF ground radial for a particular frequency, divide 234 by the number of megahertz to obtain the length in feet, or divide 71.3 by the number of megahertz to obtain the length in meters. For example, a radial for 7.030 MHz will measure 33.3 feet, or 10.1 meters, long. If you don't have a favorite frequency within a given band, use the center frequency of the band as the basis for your calculation.

Upper Floors

If you put your station on a floor above ground level, then you'll have trouble making the ground scheme of Fig. 5-3 work for an end-fed wire. You'll have to run a significant length of wire from your rig to the ground rod, and that wire will act as part of your antenna! The problem might not cause too much trouble at the lowest frequencies (1.8 MHz and 3.5 MHz), but it will affect performance at higher frequencies. If your station has to go above ground level, then, you're better off with an antenna with the feed point located a good distance away from your station. That means you'll have to use a transmission line between your radio and your antenna.

Although you can't expect to get a good RF ground from the station location on an upper floor, you can still get a good RF ground for your whole system if you put it at the feed point for, say, a ground-mounted, vertical antenna. A dipole antenna fed with a balanced open-wire line doesn't need a dedicated RF ground, nor does a Yagi or other well-designed antenna fed with coaxial cable. These antennas, in effect, supply their own RF ground at the feed point. Nevertheless, you'll always want to ensure that your rig has a substantial *electrical* ground, so that you don't get a shock from metallic parts in the event of a short circuit in the equipment somewhere.

The "Main Radio"

Some hams prefer separate transmitters and receivers, especially on the HF bands, for the "main radio." However, latter-day transceivers have gained many of the features once available only with separate units. Let's look at the basic functions and features of a typical transceiver of the sort you can get from brand-name vendors.

Overview

Fixed-station radios have power supplies that let them work directly from the 117-volt or 234-volt AC household utility mains. You can categorize these radios according to their frequency coverage. Numerous models cover 160 through 10 meters. Many, if not most, have receiving capability continuously throughout the

HF range, and some can receive as low as a few kilohertz and as high as several hundred megahertz. In a good radio, a microcomputer limits transmitting capability to the ham bands only.

Most transceivers for HF can work all popular modes including CW, SSB, AM, FM, and the various digital modes by means of audio data supplied to the microphone input when you set the rig to work on SSB. Units designed for VHF and UHF might cover one band, such as 2 meters, or two bands, such as 2 meters and 70 centimeters. A few can cover several bands. Some work only on FM, while others can work all popular modes including CW, SSB, AM, and the digital modes.

Common Features

You can acquaint yourself with all the features available in modern ham transceivers by going to a *hamfest* (organized convention of ham radio operators) and looking at radios, or by obtaining specific model brochures from manufacturers. Nearly all radios these days, especially the ones designed for HF use, have the following features:

- *Digital frequency displays* use seven-segment LED or LCD modules to give you a direct numerical readout of your frequency. While a digital display has obvious assets, a few operators prefer *analog displays* that "slide along as you tune." These days, digital displays are practically a cosmic standard, so if you're an "analog tuning addict," you might want to shop for a vintage radio at a hamfest or convention!
- A *precision readout* displays your frequency to the nearest tenth of a kilohertz, and sometimes down to a hundredth of a kilohertz (10 Hz), from 160 through 10 meters. You'll rarely see this resolution level of VHF/UHF, but some radios have it all the way from the lowest coverage frequency to the top.
- *Longwave coverage* lets you listen all the way down to 100 kHz, 30 kHz, or even 10 kHz in some radios.
- *Flywheel tuning*, in which a rotatable frequency control system acquires physical momentum when you turn it, gives the control a substantial "feel." You can rapidly carry out several revolutions of the knob with a single sharp twist; the dial keeps on going for a second or so after you let it go. Many operators like flywheel tuning, but some do not. If possible, check out some radios "live" at a hamfest or convention to evaluate them in this respect and decide whether or not you like the "feel."
- *Continuous receive coverage* allows you to listen all the way from the lowest design frequency to the highest, without any gaps.
- A *preamplifier* is a low-noise amplifier intended for boosting weak signals. It can improve the sensitivity of a receiver at the front end, but it's important that linearity be as good as possible. Nonlinearity gives rise to a wide variety of problems, especially if you use the preamplifier when some of the

signals are strong. A receiver with a built-in "preamp" should allow you to switch it off when you don't need it.

- A *noise blanker* blocks the receiver for a few milliseconds when a noise impulse comes in. This type of circuit can be extremely effective against ignition noise in some cases, but doesn't do much for "thunderstorm static" or some forms of electrical line noise. In any case, you should always look for this feature when shopping for a radio.

- *Rectangular response*—A good IF bandpass filter exhibits a so-called *rectangular response*. In theory, a rectangular response is characterized by zero attenuation inside the passband and infinite attenuation outside the passband. Look for radios that approach this ideal as closely as possible. You can figure out how well a receiver performs in this respect by looking at the *shape factor* specification (page 152) or, better yet, testing a radio in real time.

- *Band scanning*—In a channelized receiver such as you'll often find at VHF and UHF, *band scanning* can save you a lot of tedious tuning and hunting for signals. You can program the radio to constantly sweep through a given span of frequencies (the limits for which you can program into the radio's microcomputer) until it encounters an occupied channel.

- *Channel scan*—In some radios for use on any ham band from 160 meters through microwaves, you can scan through a preselected set of channels stored in memory. This mode is known as *channel scan*. Some radios can search for either a vacant frequency or an occupied frequency.

- *Programmable memory* allows you to store certain frequencies and call them up later at the push of a button, without the need for changing the band switch or VFO setting. You'll find this feature in contemporary radios for use on any of the ham bands.

- *Receiver incremental tuning* (RIT) lets you adjust the receiver frequency a few kilohertz up or down from the transmitter frequency, which remains constant as determined by the main variable frequency oscillator (VFO) or synthesizer. This feature is especially popular among contesters and DXers on the HF bands.

- *Dual VFOs* allow the receiver and transmitter frequencies to be determined independently within a given band, or, in some units, in crossband mode (transmit and receive frequencies in different amateur bands).

- *Semi break-in operation* in CW actuates the transmitter automatically when the key is closed; the receiver is disabled with a delay.

- *Full break-in operation*, also known as *full QSK*, lets you hear signals in between the dots and dashes of CW transmissions, and in some radios, between words or syllables of speech in SSB.

- *Auto-tune function* gets rid of the need to manually tune the transmitter's final amplifier when you change frequency.

- A *built-in CW keyer* lets you connect a key paddle (such as the Vibroplex Vibrokeyer, my all-time favorite!) to the radio directly, without having to purchase an outboard electronic keyer box.

Advanced Features

Sophisticated radios have capabilities that can help you with contesting, traffic handling, specialized modes, and *DXing* (seeking out contacts with stations in exotic locations).

- *Delta tuning* allows you to adjust the receiver frequency alone, or the transmitter frequency alone, for up to several kilohertz either side of the main VFO frequency. It is, in effect, incremental tuning that works with either the receiver or the transmitter. This feature is popular in DXing. Many DX station operators want you to call on a frequency other than their own transmitting frequency.
- *Panoramic reception* allows you to visually monitor a specified band of frequencies. A small screen displays the signals as vertical pips along a horizontal axis. The signal amplitude is indicated by the height of the pip. The position of the pip along the horizontal axis indicates its frequency. Usually, your frequency appears at the center of the horizontal scale.
- An *adjustable IF passband* compensates for the fact that different modes occupy different signal bandwidths. A CW receiver works well with an IF bandwidth of 500 Hz, while for SSB, an IF bandwidth of 2.7 kHz is typical. In high-noise or crowded band conditions, narrower IF bandwidth can improve the sensitivity. Advanced radios have programmable passbands, and some let you adjust the shape factor (described under "Specifications") too!
- An *automatic noise limiter* clips noise (QRN) peaks, while not affecting the desired signal. The circuit sets the clipping threshold according to the strength of the incoming signal. A limiter prevents noise from getting stronger than signals within the receiver passband. This scheme will sometimes help you hear signals through noise that defeats a standard noise blanker.
- An *audio passband filter* can enhance reception above and beyond the performance obtainable with IF filtering alone. A filter with a passband going from 300 Hz to 3000 Hz can improve the quality of reception in SSB. A CW signal needs only about 100 Hz of bandwidth in order to be clearly read at most speeds that hams use. Some CW audio filters can cut the bandwidth down to 50 Hz or less for reception at slow speeds under exceptionally adverse conditions.
- A *notch filter* is a band-rejection filter with a sharp, narrow, tunable response. By adjusting the notch frequency, you can null out interfering *heterodynes* (carriers that produce audio tones with a product detector). This feature

can prove invaluable on bands that hams must share with shortwave broadcasting, such as the upper end of 40 meters.

- *Mechanical filters* and *ceramic filters* are IF bandpass filters found in a few high-end, vintage, dual-conversion receivers with low-frequency, second IF chains, notably the old Collins Radio equipment. These filters rival today's *digital signal processing* (DSP) for performance. The input signal gets converted into electromechanical vibration by the input transducer. The resulting vibrations travel through a set of resonant disks to the output transducer. The output transducer converts the vibrations back into an electrical signal.
- *Q multiplier*—In early superheterodyne receivers, a circuit called a *Q multiplier* was occasionally used to enhance the selectivity. You'll still find such circuits in some vintage radios, along with ceramic and mechanical filters. In today's radios, DSP has pretty much taken over the IF filtering job.
- *Standing-wave ratio*—A good high-end radio will have a *wattmeter* at the antenna terminals so you can see how much RF power your transmitter puts out at any given time. Some of these meters also give you an indication of the *standing-wave ratio* (SWR) at the transmitter antenna terminals. **Caution**: Poorly designed or cheap wattmeters give accurate readings only when the SWR is 1:1 with a 50-ohm coaxial feed line (indicating a perfect impedance match between the line and the antenna).
- *Weight controls*—High-end radios with built-in electronic keyers sometimes have CW *weight controls* to compensate for the varying tastes of receiving operators, and for the effects of high speed on the signal envelope. A typical weight control allows adjustment of the dot-to-space ratio between limits of about 1:2 ("light") and 2:1 ("heavy").

Specifications

In addition to looking over the features that various radios have to offer, you'll want to know how well they do their job. You must "dig around" to get this information; product reviews (especially the ones in *QST*, the official magazine of the ARRL) can help you out. The manufacturer's technical data brochures or sheets will give you information as well.

- *Front-end specifications* get more important as the frequency gets higher, and attain paramount significance at VHF, UHF, and microwave frequencies. The front-end amplifier must be as linear as possible, and should be capable of handling very strong as well as very weak signals. If a receiver has a front end that can't handle strong signals, *desensitization* (a momentary loss of sensitivity) will occur when a strong signal comes in at the antenna terminals.
- *Noise figure* is a specification of the performance of an amplifier or receiver. This figure, expressed in decibels, tells you how much a circuit deviates from the theoretical ideal. The noise figure is important at VHF, UHF, and

microwave frequencies, where relatively little noise occurs in the external environment, and the internal noise, therefore, limits the sensitivity. Look for the lowest possible number.

- *Noise quieting* is an expression of the noise reduction in an FM receiver as a result of an incoming signal. With the squelch open and no signal, the receiver emits a hiss because of internally generated noise. When a weak signal comes in, the noise level decreases. As the signal gets stronger, the level of the noise continues to decrease. This phenomenon provides a means of measuring the sensitivity of an FM receiver. You measure the level of the noise at the speaker terminals under no-signal conditions. Then you introduce an unmodulated signal with a calibrated generator. You increase the signal strength until the noise voltage drops by 20 dB at the speaker terminals. Then you measure the signal level, in microvolts, at the antenna terminals. When shopping for a radio, look for the lowest possible microvolt figure associated with 20 dB of *noise quieting* (NQ).

- *SINAD* is an acronym based on the technical expression "signal to noise and distortion." The SINAD figure is frequently used to define the sensitivity of an FM receiver at VHF and UHF. It takes into account not only the NQ sensitivity of a receiver, but also its ability to reproduce a weak signal with minimum distortion.

- The *squelch sensitivity* of a VHF or UHF FM receiver is the signal level, in microvolts at the antenna terminals, needed to keep the squelch open continuously. Squelch sensitivity is related to noise-quieting sensitivity. The better the quieting sensitivity, the less signal it takes to actuate the squelch at the threshold. Look for the lowest possible value, expressed in microvolts at the antenna terminals.

- *Dynamic range* is the ratio, in decibels, between the levels, in microvolts, of the strongest signal that can exist at the antenna terminals without causing appreciable distortion in the front end, and the weakest signal that the radio can bring in. Most receivers today have adequate sensitivity, so that's not a primary limiting factor in the dynamic range of a receiver. Much more important is the ability of a receiver to tolerate strong input signals without developing nonlinearity. Look for the highest possible decibel figure.

- *Image rejection*—In a poorly designed receiver, you'll hear lots of phantom signals, called *images*, on frequencies where they really aren't transmitting. *Image rejection* is the extent to which images are attenuated with respect to a signal on the desired frequency. Image rejection is specified in decibels. In general, the higher the intermediate frequency of a receiver, the better the image rejection. Look for the highest possible number.

- *Intermodulation spurious-response attenuation* is expressed in decibels. It's an alternative way to express image rejection. The susceptibility or immunity of a receiver to spurious responses is primarily dependent on the design of its front end. Look for the highest possible decibel figure.

- *Harmonic suppression* is the degree to which harmonic energy is attenuated, with respect to the fundamental frequency, in the output of a radio transmitter. Emitted harmonics can cause interference to other services and operations. The FCC regulates the behavior of commercially manufactured transmitters in this respect. You'll see the figure given in decibels. Look for the highest possible number.
- *Shape factor*—In a receiver, the *shape factor* expresses how close the IF bandpass filter comes to having a rectangular response. It's determined by measuring the passband width for two attenuation levels such as −6 dB and −60 dB, and then dividing the latter number by the former. The smaller the result, the better. A perfect rectangular response would correspond to a shape factor of 1:1, or simply 1.

Peripheral Equipment

Many ham radio operators get by with a transceiver, a headset, and a microphone for their home station use. If that's all you need to start out with, that's fine, but sooner or later you'll probably want to add more stuff. Popular peripherals include *linear amplifiers*, *transmatches* (also called antenna tuners), and *digital interface* equipment.

Linear Amplifier

A linear amplifier or "linear" is an RF power amplifier that faithfully reproduces the modulation envelope of an SSB or AM signal. Linears can function with amplitude, frequency, phase, or pulse modulation. A linear will not, when operated properly, introduce significant distortion into the signal, no matter what type of emission the driving signal has.

Tip

Ideally, you should use a linear only when the power output from your transceiver is not enough to provide reliable communications. Extra power amplification should not be used when it's not necessary. That practice (all too common, unfortunately) is not only wasteful, but technically, it's illegal! Specific FCC regulations dictate that hams should not use more power than they need in order to carry out the desired communications. Lawyers might "split hairs" on this issue, but you can certainly use a bit of good sense here. If someone tells you you're "20 over 9" (extremely loud signal) while you're running 1500 watts output in a casual contact, you might as well power your linear down and run "barefoot."

Linears have always been popular among DXers and contesters. They have also proven valuable in traffic handling, especially when the traffic has a priority or emergency nature, and you want to ensure that your messages reach their destinations in the shortest time possible, without having to repeat yourself because of interference or marginal band conditions.

A commercially manufactured linear usually comes from the vendor as a self-contained unit, sometimes with a built-in power supply and sometimes with an outboard supply. The most rugged linears can provide 1500 watts continuous (100-percent duty cycle) RF output, the maximum legal limit for amateur service. Some linears are designed to supply 1500 watts output for a 50-percent duty cycle, typical CW and SSB. Other linears can't provide the maximum legal limit but instead go with something less, such as 700 watts or 1000 watts. These units might work for hams on a strict budget, or who do not have access to the recommended 234-volt AC utility panel that a full-power linear should have.

Today's linears have features that make them versatile indeed compared to their "boat anchor" ancestors! These attributes can include built-in wattmeters, built-in *SWR meters*, provisions for full CW break-in option, automatic shutdown in case of overheating, *automatic gain control* (AGC), and automatic tune-up or broadband operation.

Did You Know?

A typical linear needs 70 to 100 watts of drive power to produce its maximum power output. A few linears can work off of much less drive, but they usually offer something less than the maximum legal limit of RF power output.

For Nerds Only

The term *linear* means that the output envelope has the exact same shape as the input envelope does. In engineering terms, the amplification curve (instantaneous output versus instantaneous input) shows up as a straight line on a graph. In other words, the instantaneous output is a *linear function* of the instantaneous input.

Tip

You can find detailed information about the various classes of RF power amplifiers in my book *Teach Yourself Electricity and Electronics*. Look for the latest edition. I offer a link to all my books from my website at

www.sciencewriter.net

Barring some unforeseen media meltdown, you'll also find free videos on these topics if you visit and browse through my YouTube channel, also linked through my website.

Figure 5-4 is a simplified schematic diagram of a scaled-down, broadband linear that operates with a single power transistor. Larger linears (providing more power) have essentially this same configuration, but might need two or three large power transistors connected in parallel.

Transmatch

The output impedance of a radio transmitter, or the input impedance of a receiver, should ideally be matched to the antenna-system impedance. This state of affairs requires that the antenna system present a *nonreactive load* (no capacitance or inductance, but only the esoteric phenomenon known as *radiation resistance*) of a certain value, usually 50 ohms. Very few antenna systems meet this requirement to perfection, but you can match almost any load impedance to your radio by means of a transmatch, which comprises inductors and capacitors that cancel out the antenna's reactance and convert the remaining resistance to the appropriate value to provide a match to the coaxial cable.

Figure 5-5 shows four simple transmatch circuits. The arrangements at A and B work with unbalanced loads, such as vertical and ground-plane antennas; the circuits at C and D work with balanced loads, such as dipole antennas and large loops. Most transmatch circuits incorporate SWR meters (reflectometers) in their input circuits to facilitate adjustment. Some such meters actually show you the forward and reflected power in watts. Some reflectometers have two separate needles and three scales. One scale and needle indicate forward power; a second scale and needle indicate reflected power; a third scale, calibrated according to the crossing point of the two needles, indicates the SWR.

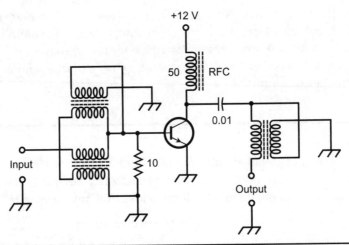

FIGURE 5-4   A broadband RF power amplifier, capable of producing a few watts of output. Resistances are in ohms. Capacitances are in microfarads (μF). Inductances are in microhenrys (μH). The 50-μH component labeled "RFC" is an RF choke.

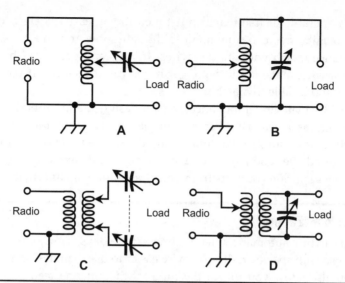

FIGURE 5-5 Simple transmatch (antenna tuner) circuits. At **A**, a tuner for low-impedance unbalanced loads; at **B**, a tuner for high-impedance unbalanced loads; at **C**, a tuner for low-impedance balanced loads; at **D**, a tuner for high-impedance balanced loads.

If you use a transmatch with your radio, you should make preliminary adjustments using an *impedance bridge* if you can get your hands on one. This little gadget eliminates the need for transmitting signals over the airwaves. If you can't find an impedance bridge, you can adjust a manual transmatch according to the following procedure.

- Set your receiver and transmitter to the same frequency.
- Adjust the transmatch until the signals and/or noise "peak" in the receiver.
- Adjust the transmitter tuning and loading (if your transmitter has such controls) to the approximate settings for the frequency that you want to use.
- Switch the transmitter on, and apply a low-power signal to the transmatch.
- Adjust the transmatch for minimum SWR as indicated on a reflectometer placed between the transmitter and the transmatch.
- Retune the transmitter for normal operating output power. In latter-day radios, this process may involve nothing more than hitting a "tune" button for a couple of seconds.
- Identify your station.
- Record the positions of the transmatch controls for future reference.

Transmatches are especially handy for portable station operation. You can string up a random-wire antenna in any fashion and tune it to resonance by using a transmatch, unless it happens to be close to an exact multiple of 1/2 wavelength long. Generally, such antennas must measure at least 1/4 electrical wavelength if good results are to be obtained, although some transmatches can provide an impedance match with shorter wires. If you can't get a match with a particular random wire, make it a little longer if you can (or shorter if you must), and then you should be able to match it.

Besides matching impedances, a well-designed and properly operated antenna tuner acts as a *lowpass filter* or *bandpass filter*, providing additional front-end selectivity for your receiver, improving the image rejection, and reducing the risk of front-end overload in case of a strong signal off-frequency. The transmatch also attenuates spurious outputs from your transmitter, including the harmonics.

Here's a "Zapper"!

You can buy automatic antenna tuners these days, some of which are designed for use with specific radios. They're easier to use than the old manual ones, and they also allow you to put the tuner at the antenna feed point, where it can ensure a 1:1 SWR along the entire length of your feed line. However, some commercially manufactured automatic antenna tuners have sensitive internal electronics that can suffer destruction from the strong EM field produced by a nearby lightning strike. The EM field induces a *current surge* that flows along the control line between the radio and the transmatch. A surge of this sort "zapped" my feed-point tuner on a summer afternoon a few years ago, even though I had disconnected and grounded the antenna. Since that time, I haven't replaced that tuner for fear that the same thing will happen again. Instead, I rely on a manual tuner with large, discrete, rugged internal inductors and capacitors, even though it's less convenient. It should survive anything short of a direct hit!

Digital Interface Equipment

If you want to take advantage of the fascinating non-CW digital modes that exist today, you'll need to equip your rig with a digital interface. At the very least, you'll need a computer, although it doesn't have to be a powerful one. An old computer, defunct for most other applications, will usually work fine for RTTY, PSK, MFSK, and other emerging digital modes. You should, in any case, have a good display so that you don't get eyestrain from looking at all that decoded text!

In order to receive signals on any non-CW digital mode, you need to connect the audio output of your radio (from the headset jack, in most cases) to the "line in" jack on your computer. (If you use the microphone input jack on your computer, you'll probably get too much audio, with consequent distortion.) Set the level on the sound card of your computer for a comfortable volume at the computer headset jack or speakers when your radio's audio gain is set close to the middle of its range.

Once you've gotten the audio from your radio into your computer, you'll need to download a program (or two, or three) that can decode the signals in the digital mode of interest. I use a program called *DigiPan* for PSK, another program called *HamScope* for PSK and MFSK, and a program called *MMTTY* for RTTY. All three of these programs are available online for free, and they all work on most versions of the Microsoft operating system (OS). Other programs exist, but most of them require payment. None of the ham radio digital programs take up much computer processing power or memory, so they run fast even on simple machines. Enter the name of the program of interest into your favorite search engine, and follow the links until you get a download that works.

Heads Up!

Before you download any program onto any computer, create an operating system restore point. That way, if the program doesn't work, you can (in addition to deleting the program itself) take your OS back to the state it was in before you downloaded the program. You should also scan your computer with an up-to-date, anti-malware utility after you've downloaded any program, whether it works or not.

In order to use a program once you've downloaded it, I recommend that you play around with it and see which button produces what action! (To heck with "help" menus!) I have a personal paradigm for learning freeware programs such as the ones mentioned above: COIES ("click on everything in sight"). You might expand this concept to COEISASWH ("click on everything in sight and see what happens"). If you conduct that exercise for a good long while, you'll know how to use the program well enough so that you're ready for the next step: Getting a digital interface unit. That's a little box about the size of an old-fashioned outboard electronic keyer. Visit a ham radio convention or your local ham radio club and "pick some brains" for specific interface recommendations.

An Old Timer Remembers

In 2004 when I first got interested in PSK, I bought an interface box called the *RigBlaster Pro* from *West Mountain Radio*. That company still exists, and has expanded their line of products. That thing has worked flawlessly for more than a decade. Point your Web browser to

http://www.westmountainradio.com

Besides the RigBlaster Pro (which they still offer), they have simpler digital interface units for hams with limited objectives. The RigBlasterPro allows you to easily transmit, as well as receive, signals in PSK and numerous other digital modes with the appropriate software. West Mountain Radio, at the time of this writing (2014), offers software for more than 20 exotic digital modes!

Utility-Operated Power Supplies

A fixed-station power supply provides your home station with the proper voltages and currents from the AC utility line. Nearly all active electronic devices require DC. Solid-state equipment commonly operates from 6 volts to 12 volts DC. Ham equipment using vacuum tubes needs 100 volts to 3000 volts DC. Here's a breakdown of the components in a typical power supply of the sort used with fixed-station ham radio transceivers these days, along with a little "inside information" about how they work.

Power Transformers

We can categorize *power transformers* in two general ways: *step-down* or *step-up*. The output, or secondary, voltage of a step-down transformer is lower than the input, or primary, voltage. The reverse holds true for a step-up transformer, in which the output voltage exceeds the input voltage.

Most latter-day electronic devices need only a few volts to function. The power supplies for such equipment use step-down power transformers, with the primary windings connected to the utility AC outlets. The transformer's physical size and mass depend on the amount of current that you expect it to deliver. Some devices need only a small current at a low voltage. The transformer in a radio receiver, for example, can be physically small. A large Amateur Radio transmitter or linear amplifier needs more current. The secondary windings of a transformer intended for that application must consist of heavy-gauge wire, and the cores must have enough bulk to contain the large amounts of magnetic flux that the coils generate.

Some circuits need high voltage. The cathode-ray tube (CRT) in an old-fashioned home television (TV) set needs several hundred volts, for example. Some Amateur Radio power amplifiers use vacuum tubes, even today, which need upwards of 1000 volts! The transformers in these linears are step-up types. They must have considerable bulk because of the number of turns in the secondary, and also because high voltages can spark, or arc, between wire turns if the windings aren't spaced far enough apart.

Rectifier Diodes

Rectifier diodes are available in various sizes, intended for different purposes. Most rectifier diodes are made from silicon, so engineers call them *silicon rectifiers*. Some rectifier diodes are fabricated from selenium, so engineers call them *selenium rectifiers*. When you work with power-supply diodes, you must pay close attention to two specifications: the *average forward current* (I_o) rating and the *peak inverse voltage* (PIV) rating.

If you drive too much current through a diode, the resulting heat will destroy it. When designing a power supply, you must use diodes with an I_o rating of at least 1.5 times the expected average DC forward current. If this current is 4 amps, for example, the rectifier diodes should be rated at $I_o = 6$ amps or more.

The PIV rating of a diode tells you the maximum instantaneous *reverse-bias voltage* that it can withstand. A well-designed power supply has diodes whose PIV ratings significantly exceed the peak AC input voltage. If the PIV rating is not great enough, the diode or diodes in a supply will conduct current for part of the reverse cycle, degrading the efficiency.

Half-Wave Circuit

The simplest rectifier circuit, the *half-wave rectifier* (Fig. 5-6A), has a single diode that "chops off" half of the AC cycle. This method of rectification has shortcomings. First, the output is difficult to filter; that is to say, you'll find it hard to get rid of the pulsations (called *ripple*) in the DC that comes out of the diode. Second, the output voltage can drop considerably when the supply must deliver high current. Third, half-wave rectification puts a strain on the transformer and diodes because it works these components hard during half the AC cycle and lets them "loaf" during the other half.

Full-Wave Center-Tap Circuit

You can take advantage of both halves of the AC cycle by means of *full-wave rectification*. A *full-wave, center-tap rectifier* has a transformer with a connection called a *tap* at the center of the secondary winding (Fig. 5-6B). The tap connects directly to electrical ground. This arrangement produces voltages and currents at the ends of

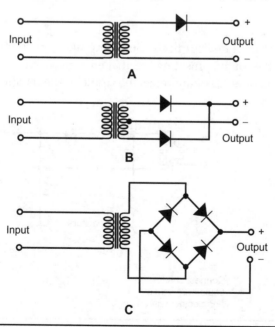

FIGURE 5-6 At **A**, a half-wave rectifier circuit; at **B**, a full-wave center-tap rectifier circuit; at **C**, a full-wave bridge rectifier circuit.

the secondary winding that oppose each other in phase. These two AC waves are individually half-wave rectified.

Full-Wave Bridge Circuit

You can get full-wave rectification using a circuit known as a *full-wave bridge rectifier*. Figure 5-6C shows a schematic diagram of a typical full-wave bridge circuit. It does not require a center-tapped transformer secondary, so you can get away with a less expensive transformer. In addition, this method of rectification uses the entire secondary winding on both halves of the wave cycle, so it makes more efficient use of the transformer than the center-tap circuit does. The bridge circuit also places less strain on the individual diodes than a half-wave or full-wave center-tap circuit does.

Voltage-Doubler Circuit

You can connect diodes and capacitors together to deliver a DC output of approximately twice the positive or negative peak AC input voltage. Engineers call this arrangement a *voltage-doubler power supply*. It's a compromise, though, if you want to get high DC voltage, because voltage drops considerably when you demand a lot of current from it. If the current fluctuates, so does the output voltage (the higher the current, the lower the voltage). Nevertheless, in low-current systems, this type of rectifier can work okay. Figure 5-7 is a simplified diagram of a voltage-doubler power supply. It takes advantage of the entire AC cycle, so it constitutes a *full-wave voltage doubler*. It serves as its own filter because the capacitors "smooth out" the pulsations in the DC that comes out of the diodes.

Filtering

Most DC-powered devices need something like the DC that would come from a battery, not the rough, pulsating DC that emerges from most simple rectifiers. You can eliminate, or at least minimize, the ripple in the rectifier output using a *power-supply filter*.

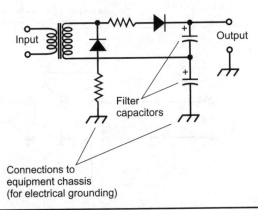

FIGURE 5-7   A full-wave, voltage-doubler power supply.

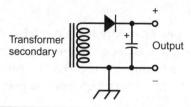

Figure 5-8   You can use a large-value capacitor all by itself as a power-supply filter.

The simplest power-supply filter consists of one or more large-value capacitors, connected in parallel with the rectifier output, as shown in Fig. 5-8. An electrolytic capacitor makes a good component for this purpose. It's *polarized*, meaning that you must connect it in the correct direction. A given electrolytic capacitor also has a certain maximum rated working voltage. Pay attention to these particulars!

You can obtain enhanced *ripple suppression* by placing a large-value inductor in series with the rectifier output along with a large-value capacitor in parallel. When an inductor serves in this role, engineers call it a *filter choke*. In a filter that uses a capacitor and an inductor, you can place the capacitor on the rectifier side of the choke to construct a *capacitor-input filter* (Fig. 5-9A). If you locate the filter choke on the rectifier side of the capacitor, then you have a *choke-input filter* (Fig. 5-9B).

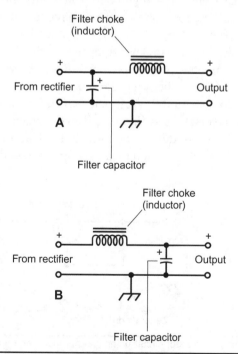

Figure 5-9   At **A**, a capacitor-input filter; at **B**, a choke-input filter.

Voltage Regulation

When you expect a power supply to deliver high current intermittently, you can employ a *power transistor* along with a *Zener diode* to obtain voltage regulation, using a circuit such as the one in Fig. 5-10. You can also find voltage regulators in *integrated-circuit* (IC) form. Such a *regulator chip* goes in the power-supply circuit at the output of the filter.

"Tanglewire Gardens"

Most of us have computer workstations, usually with multiple peripherals and ancillary equipment, such as a printer, a scanner, a modem, a router, a cordless phone, a desk lamp or two, a charging bay (for devices such as tablet computers and cell phones), and so on. All of these things get their power, either directly or indirectly, from the 117-volt utility system. As a result, anyone with a substantial computer workstation will end up with a "tanglewire garden" behind and under the work desk. The same thing will happen eventually, if you gather enough gear in your ham shack.

"Tanglewire gardens" can look dangerous, as if they would present a high fire risk, but they needn't pose a hazard. If you know how to connect and arrange the wires properly, it doesn't matter from a safety standpoint how much you snarl them up, although you might want to affix labels on the cords near their end connectors (on each end) so that you don't get them confused with each other when the inevitable malfunction occurs and you have to pull out and replace one of the components of your system.

Figure 5-11 shows the "tanglewire garden" underneath my ham radio station. In addition to the radio itself, this system includes a computer, two displays, a digital communications interface between the radio and the computer, a microcomputer-controlled reflectometer and RF wattmeter, an audio amplifier for the computer and radio, a wireless headset, a desk lamp, and an external hard drive that needs its own "power brick." That's 10 devices or cords in total, all deriving their power from a single outlet in the wall protected by a 15-amp breaker at the main utility box.

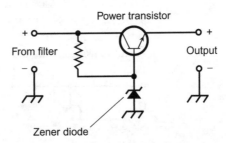

FIGURE 5-10   A voltage-regulator circuit using a Zener diode and a transistor.

FIGURE 5-11   "Tanglewire garden" beneath the author's electronics workbench. A heavy-duty UPS (out of the picture to the right) serves two power strips mounted on a metal baking sheet that rests on detached plastic shelves.

In order to ensure smooth operation of the system in case of a power failure, all of the devices go to the wall outlet through a commercially manufactured *uninterruptible power supply* (UPS). The UPS has a battery that charges from utility electricity under normal conditions, but provides a few minutes of emergency AC (with the help of some sophisticated electronic circuits) if the utility power fails. That few minutes gives me time to deploy my backup generator, described later in this chapter, without having to shut any of the devices down. The UPS has four outlets in the back, two of which go to power strips with six outlets each, and the other two of which remain empty. There are 12 available outlets in the power strips, 10 of which are in active use. The UPS also has a transient suppressor built-in. Figure 5-12 is a block diagram of the general arrangement.

Did You Know?

You should not use power strips with transient suppressors in conjunction with any other component, such as a UPS, that also has a transient suppressor. When you cascade transient suppressors, they'll likely conflict. That's why, in the system shown by Figs. 5-11 and 5-12, the power strips do not include transient suppressors. They do, however, each have 15-amp circuit breakers for extra protection in case of an equipment short.

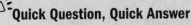

Quick Question, Quick Answer

- Doesn't the presence of 10 devices, all plugged into a single wall outlet, as shown in Figs. 5-11 and 5-12, create a danger by overloading the wall outlet and its associated house wiring?
- Not in this case! All of the devices, taken together, consume less than 10 amps (2/3 of the breaker rating), even if they all run at once. The radio interface, the cordless headset, the audio amplifier, the power-measuring meter, and the desk lamp draw less than 1 amp combined. The rest of the devices, taken together, draw about 7 amps.

I've taken three extra precautions, aside from making sure that I don't overload the wall outlet, to ensure that my "tanglewire garden" remains safe. You should do the same with your fixed station equipment!

1.   First, if you look carefully at Fig. 5-11, you'll notice that I've mounted the power strips on a metal sheet. It's a solid aluminum baking sheet. I glued the strips down there with epoxy resin. This precaution keeps the power strips from setting anything (other than themselves) on fire if they start shorting out and sparking, a stunt that these things have been known to perform, occasionally with disastrous results.

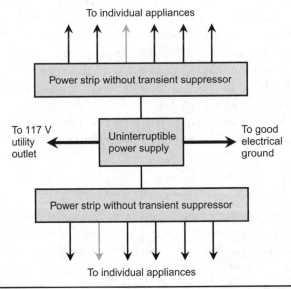

Figure 5-12   Block diagram of the "tanglewire garden" beneath the author's workbench. The power strips include breakers but not transient suppressors; the UPS contains a transient suppressor that serves the whole system. Gray arrows represent unused outlets.

2.   Second, I don't let any cord splices or other sensitive electrical points lie directly on the floor. The baking sheets, as well as all points in the cords where splices exist, are set up on thick plastic shelves. Although I've never been flooded out, my basement floor will get wet if a sudden cloudburst occurs. (Of course, in that event I won't power up the ham shack at all until the floor dries out!)

3.   Third, I've connected a dedicated ground wire from the chassis of the UPS to a known electrical ground. I tested the wall outlet underneath the workbench to ensure that the "third prong" actually goes to the electrical ground for the house.

Small Backup Generators

You can find compact, portable *combustion generators* for use in homes and small businesses. Some combustion generators are also suitable for use by campers. For people living in remote areas, a combustion generator might constitute the primary, if not the only, source of AC electricity for common appliances. And of course, every good fixed ham radio installation ought to have one!

How They Work

A small combustion generator provides 117 volts AC in the United States (234 volts in many other countries) at 60 Hz. Larger generators in the United States also supply 234 volts AC at 60 Hz for heavy appliances, such as electric ranges and laundry machines. The generator's internal combustion engine can range in size from a few horsepower, comparable to the one in a lawn mower or snow blower, to hundreds of horsepower, comparable to the engines in trucks, tractors, and construction equipment.

> **Tip**
> Most small generator engines burn gasoline. Larger ones burn diesel fuel, propane, or methane.

In the simplest type of AC generator, a coil of wire, attached to the shaft of the combustion engine, rotates between a pair of powerful magnets. If you connect a load, such as your radio, to this coil, an AC voltage appears across that load as each point in the wire coil moves past the *lines of flux* produced by the magnets, first in one direction and then in the other direction, over and over. In an alternative arrangement, the magnetic poles revolve around the wire coil, which remains fixed.

The AC voltage that a generator can produce depends on the strength of the magnets, the number of turns in the wire coil, and the speed of rotation. The AC frequency in a simple generator depends only on the speed of rotation. In the United States, the speed is 3600 revolutions per minute (3600 r/min) or 60 revolutions per second (60 r/s), resulting in an output frequency of 60 cycles per second (60 Hz). In many other countries, the rotational speed is 3000 r/min, producing an AC frequency of 50 Hz. In order to maintain a constant rotational speed for the generator under conditions of variable engine speed, mechanical regulating devices are required.

When you connect a load to the output of a simple generator, the engine has a harder time turning the generator shaft, as compared with the situation when no load exists. As the amount of electrical power demanded from a generator increases, so does the mechanical power required to drive it, and therefore, the amount of fuel consumed per unit of time. The electrical power that comes out of a generator is always less than the mechanical power required to drive it. The lost energy shows up as heat in the generator components. To maintain the proper AC frequency, a simple generator's engine must run at a constant speed under conditions of variable load. That state of affairs can prove difficult to attain, but latter-day engineers have found a way around the trouble!

For Nerds Only

The *efficiency* of a generator equals the ratio of the electrical power output to the mechanical driving power, both measured in the same units, such as watts or kilowatts, multiplied by 100 to get a percentage. No generator operates at 100-percent efficiency, but a good one can come fairly close to that ideal.

Small Gasoline-Powered Generators

Advanced small-scale generators circumvent the need for constant motor speed by converting the generated AC to regulated DC, and then using a *power inverter* to generate AC from that DC. If the motor speed changes, the DC voltage stays the same because the regulator circuit holds it constant, so the output AC voltage stays constant too. In the best commercially manufactured generators, the inverter produces a near-perfect sine wave to ensure that the machine can properly operate sensitive electronic devices, such as computers, computer-controlled radios, printers, scanners, modems, and routers.

Figure 5-13 shows my portable gasoline generator with a power inverter that can provide around 1600 watts of clean sine-wave AC electricity when needed. This machine can run my ham station under full-power-transmit conditions, as well as all three of my computers and the microcomputer-controlled furnace, all at once. It has a tank that holds 1.1 gallons (4 liters) of high-octane, unleaded gasoline. With a load of a few hundred watts, that amount of fuel provides several hours of continuous, reliable AC electricity. This generator has proven itself worthy as a backup power source in winter storms when utility failures would otherwise have meant no heat for my house, as the furnace electronics and fan require 117 volts AC to function!

Did You Know?

Any backup generator, if poorly designed, can cause problems if you try to run sensitive electronic equipment from it. However, a well-engineered generator with a power inverter, even the small gasoline-fueled type, will work fine with computers, microcomputer-controlled ham radios, and other sophisticated systems, as long as you keep it in proper working order and don't overload it. If you want your generator to work when you need it, you must adhere to a maintenance schedule that involves cleaning, spark-plug replacement, and periodic testing.

My Arrangement

The Honda EU-2000i portable gasoline-fueled generator (Fig. 5-13) forms the heart of my emergency backup power ensemble. In addition to the generator, I use several extension cords and power strips to distribute electricity to the points where I need it the most during a utility outage. I always keep in mind the maximum power that the generator can provide; I never let it come close to "maxing out." You can use this general configuration as a template for your own system if you want to install one, tailoring the specifics to meet your needs.

FIGURE 5-13 A portable, gasoline-fueled generator, capable of providing up to 1600 watts of clean sine-wave AC power at 117 volts. (The tied-up cord is the ground wire.)

Did You Know?

Honda, Yamaha, and other manufacturers offer several portable generator models, some of which are a little smaller than mine, and some of which are quite a lot bigger. Whatever brand of generator you decide to buy, you should make certain that it produces a "clean" sine wave for sophisticated electrical and electronic devices. Always check not only with the generator dealer (who will tell you the truth in a perfect world, but in the real world, maybe not), but also with the manufacturer's specification sheet. Contact the manufacturer and ask for details about the model that interests you. Remember that the best generators employ power inverters to produce clean, regulated, sine-wave AC. Don't scrimp on this investment!

Figure 5-14 shows the two AC outlets on my generator. The cord on the left goes to the furnace fan and electronics. The cord on the right goes to my ham shack, several LED lights for the house, and all my computers by way of a power strip in the garage with an LED lamp to indicate when it's getting generator power. (This strip *does not* have a transient suppressor because the UPS, which also goes to all the electronics that I want to provide with backup power, has a built-in transient suppressor.) When a power failure occurs, I unplug the UPS from the utility mains and plug it into the generator, physically removing the plug from the utility outlet before putting it into the generator outlet.

Always try to balance the loads among multiple outlets in a generator that has more than one outlet. Ideally, each outlet should do roughly the same amount of work. This precaution ensures that the generator will operate at maximum efficiency. In some generators, the outputs appear in different phases. If I were to connect the entire load to, say, the left-hand outlet in the situation of Fig. 5-14, it would be like seating all the passengers on the left-hand side of an aircraft. The generator would function, but probably not at its best efficiency.

As a final precaution to keep the generator operating at its best, you should connect the generator's ground terminal to a known electrical ground that you have tested for continuity with the main ground for your whole house. My arrangement comprises a single heavy length of wire and a clamp going to a cold water pipe. I've satisfied myself that the cold water pipe connects directly to the main electrical ground for the house by performing a continuity test.

Warning!   Always locate a generator so that its exhaust can vent freely to the outside. The best way to make that happen is to keep the generator outdoors when running it. Never run your generator in a garage (even an open one) or a partially enclosed space of any kind. Buy a carbon-monoxide detector if you don't

FIGURE 5-14   My portable generator has two AC outlets. The cord on the left goes to the furnace electronics and fan; the cord on the right goes to the ham shack.

already have one, and place it in your house near the rooms where you sleep. Keep its batteries fresh. That way, you'll know if generator exhaust "blows" into the house, a situation that can arise with amazing ease, as I discovered when I ran it in the woodshed under my dining room. My carbon-monoxide detector sounded its alarm after only a few minutes of generator run time!

Warning!   An on-site generator must run only when your house wiring is completely separated from the electric utility wiring with a *double-pole, double-throw* (DPDT) *isolation switch* installed and tested by a certified electrician. Alternatively, you can plug appliances into the generator through cords that have *nothing whatsoever* to do with your house wiring, as I do. If you don't follow these rules strictly, *backfeed* can occur, in which electricity from the generator gets into the utility lines near the home or business where the generator operates. Backfeed can endanger utility workers and damage electrical system components.

And in case you've wondered, switching off the main breaker at your utility box *will not guarantee* that no stray voltage can make its way onto the power lines outside your house. Avoid backfeed at all costs!

Noise, Noise, Noise!

In wireless communications practice, RF noise that comes from outside is called *external noise*. The more sensitive the receiving equipment, and the longer the distance over which it has to work, the more significant this type of noise becomes.

Cosmic Noise

Electromagnetic radiation from outer space, called *cosmic noise*, occurs throughout the entire EM spectrum, from the VLF radio band where waves measure tens of kilometers long to the realm of X rays and gamma rays where scientists measure wavelengths in nanometers (millionths of a millimeter) and picometers (thousandths of a nanometer). At the low radio frequencies, the ionized upper atmosphere of our planet prevents the noise from reaching the surface. At some higher radio frequencies, the lower atmosphere prevents the noise from reaching us. But at many frequencies, cosmic noise arrives at the surface at full strength.

Cosmic noise can be identified by the fact that it correlates with the plane of the *Milky Way*, our galaxy. The strongest *galactic noise* comes from the direction of the constellation *Sagittarius* ("The Archer") because this part of the sky lies on a line between our Solar System and the center of the galaxy. Galactic noise was first noticed and identified by *Karl Jansky*, a physicist working for the Bell Laboratories in the 1930s. Jansky conducted experiments to investigate and quantify the earth's atmospheric noise at a wavelength quite close to that of the ham radio 21-MHz band. He found some radio noise that he couldn't account for, and then he noticed that its orientation correlated with the location of the Milky Way in the sky. Jansky's antenna was a simple affair like the ones used by hams.

Along with noise from the sun, the planet Jupiter, and a few other celestial objects, galactic noise accounts for most of the cosmic noise arriving at the surface of the earth. Other galaxies radiate noise, but because those external galaxies lie much farther away from us than the center of our own galaxy does, sophisticated equipment is needed to detect the noise from them.

Did You Know?
Radio astronomers analyze cosmic noise in an effort to improve their understanding of the universe. To them, humanmade noise and interference (including legitimate broadcast and communications signals) constitutes a major nuisance, just as it does with ham and shortwave radio enthusiasts!

Here's a Tale!

In 1965, *Arno Penzias* and *Robert Wilson* of the Bell Laboratories observed cosmic noise that seemed to come from everywhere, from all directions at once. For some time, the noise source remained a mystery. Nowadays, most astronomers believe that the noise originated with the fiery birth of our universe (an event often called the *Big Bang*), and comes to us "delayed" by billions of years! If they're right, then when we detect and record this noise, we in effect "hear" the echo of Creation.

Solar Noise

The amount of radio noise emitted by the sun is called the *solar radio-noise flux*, or simply the *solar flux*. The solar flux varies with frequency. But no matter what the frequency (or wavelength), the level of solar flux increases just after a solar flare occurs. A sudden increase in the solar flux indicates that shortwave broadcasting and communications conditions (including ham radio) might deteriorate within a few hours.

The solar flux is commonly monitored at a wavelength of 10.7 centimeters, which corresponds to a frequency of 2800 MHz. At this frequency, which is about 100 times the frequency of the ham radio 10-meter band, the earth's atmosphere has little or no effect on radio waves, so the energy reaches the surface at full strength.

For Nerds Only

The 2800-MHz solar flux correlates with the 11-year *sunspot cycle*. On the average, the solar flux is highest near the peak of the sunspot cycle, and lowest near a sunspot minimum. As of this writing, you can read data about solar activity by visiting the website of the American Radio Relay League at

www.arrl.org

Look for a link dealing with the "solar update."

Atmospheric Noise

Electromagnetic noise is generated in the atmosphere of our planet, mostly by lightning discharges in thundershowers. This noise is called *sferics*. In a radio receiver, sferics cause a faint background hiss or roar, punctuated by bursts of sound we call "static." Figure 5-15 shows an example of this noise as it would look on the display of a laboratory oscilloscope connected into a radio receiver.

A gigantic voltage constantly exists between the surface of the earth and the ionosphere. The earth and ionosphere behave like concentric, spherical surfaces of a massive capacitor, with the troposphere and stratosphere serving as the insulating

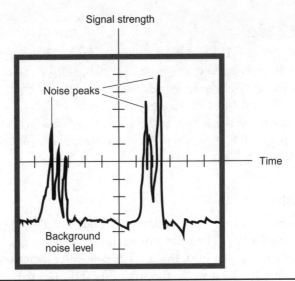

FIGURE 5-15   If you connect an oscilloscope into a radio receiver and listen to sferics, you'll see a display that looks something like this drawing.

material (called a *dielectric*) that keeps the charges separated. Sometimes this dielectric develops "holes," or pockets of imperfection, where discharge takes place. Such "holes" are usually associated with thundershowers. Sand storms, dust storms, and volcanic eruptions also produce some lightning, contributing to the overall sferics level.

Sferics are not confined to our planet! A great deal of radio noise is generated by storms in the atmosphere of the planet Jupiter. Astronomers can "hear" this noise with radio telescopes. Sferics probably also occur on Saturn, and perhaps on Uranus, Neptune, Venus, and Mars as well. In the cases of Venus and Mars, dust storms and volcanic eruptions would likely be the cause of sferics.

For Nerds Only

You can hear the sferics from a distant thundershower on the standard AM broadcast band. If you have a general-coverage shortwave communications receiver, you can listen at progressively higher frequencies as the storm or storm system approaches. When you hear the static bursts at 30 MHz, you can have confidence that the storm has come to within 100 miles (160 kilometers) of your station.

Precipitation Noise

Precipitation noise, also called *precipitation static*, is radio interference caused by electrically charged water droplets or ice crystals as they strike metallic objects, especially antennas. The resulting discharge produces wideband noise that sounds similar to the noise generated by electric motors, fluorescent lights, or other appliances.

Precipitation static is often observed in aircraft flying through clouds containing rain, snow, or sleet. But occasionally, precipitation static occurs in radio communications installations. This phenomenon is especially likely to happen during snow showers or storms; then the noise is called *snow static*. Precipitation static can make radio reception difficult, especially at low frequencies (long wavelengths).

In a good radio receiver, a *noise blanker* or *noise limiter* can reduce the interference caused by precipitation static. A means of facilitating electrical discharge from an antenna, such as a large-value, heavy-duty inductor between the antenna and ground, can also help. If the antenna elements have sharp points on the ends, you can blunt them by installing small metal spheres on those ends. If you do that, you'll have to shorten the elements slightly to make up for the *loading effects* that the spheres introduce.

Coronae

When the voltage on an electrical conductor (such as an antenna or high-voltage transmission line) exceeds a certain level, the air around the conductor begins to ionize. The atoms in the air gain or lose electrons, so that they become electrically charged. If the effect becomes significant, it can cause a blue or purple glow called a *corona* that can be seen at night. This glow commonly occurs at the ends of a radio-broadcast or communications-transmitting antenna element when the transmitter has high power output. Coronae occur increasingly often as the relative humidity rises because it takes less voltage to ionize moist air than it takes to ionize dry air. Coronae produce a strong hiss or roar in radio receivers.

Coronae can occur inside a coaxial cable that feeds an antenna just before the dielectric material breaks down, ruining the cable for good. Poorly designed antennas with high-power transmitters can subject a transmission line to that sort of stress. So can nearby thundershowers. A corona is sometimes observed between the plates of capacitors handling large voltages. Also it is more likely to occur at the end of a pointed object, such as the end of a whip antenna, than on a flat or blunt surface. Some antennas have small metal spheres at the ends to minimize the occurance of coronae.

Did You Know?

Coronae sometimes occur as a result of high voltages caused by static electricity during thunderstorms. Such a display is occasionally seen at the tip of the mast of a sailing ship. Coronae were observed by seafaring explorers hundreds of years ago. They called it *Saint Elmo's fire*, and, not knowing about electricity, some of them attributed it to supernatural forces.

Impulse Noise

Any sudden, high-amplitude voltage pulse will generate an RF field, and radio receivers will often pick it up. It's called *impulse noise*, and it can come from all kinds of household appliances, such as vacuum cleaners, hair dryers, electric blankets, thermostats, and fluorescent-light starters. Impulse noise tends to get worse as the frequency goes down, and can plague AM broadcast receivers to the consternation of their users. Serious interference can occur in shortwave receivers, but it gets less severe as the frequency rises, and it rarely poses a problem above 30 MHz.

You can minimize impulse noise problems by ensuring that you have a good ground system. All the components in the system should be grounded by individual wires to a single point. A noise blanker or noise limiter can also help, if the receiver has one. A shortwave or ham radio receiver should be set for the narrowest response bandwidth consistent with the mode of reception.

Ignition Noise

Ignition noise is impulse noise generated by the electric arcs in the spark plugs of an internal combustion engine. Many different kinds of devices produce it, including automobiles and trucks, lawn mowers, and gasoline-engine-driven generators. Figure 5-16 shows how ignition noise would look on an oscilloscope connected into the sensitive amplifiers of a radio receiver.

In a shortwave or ham radio receiver, a noise blanker can often work wonders to get rid of ignition noise problems. (In stubborn cases, the noise blanker can't deal

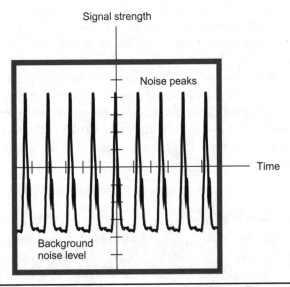

FIGURE 5-16   Here's what ignition or impulse noise looks like on an oscilloscope display connected into a radio receiver.

with the energy pulses, and you have to rely on a limiter instead.) The pulses of ignition noise are of very short duration, although their peak intensity can be considerable. Noise blankers are designed to literally switch the receiver off during these brief spikes.

Power-Line Noise

Utility lines, in addition to carrying the 60-Hz AC that they're meant to transmit, carry other currents. These currents have a broadband nature. They result in an effect called *power-line noise*. The "rogue currents" usually occur because of electric *arcing* at some point in the utility grid. The arcing might originate in household appliances; it can take place in faulty or obstructed utility transformers; it can occur in high-tension lines as a corona discharge into humid air. The currents cause the power line to radiate EM fields like huge radio transmitting antennas!

Power-line noise sounds like a buzz, hiss, or roar when picked up by a radio receiver. Some types of power-line noise can be attenuated by means of a noise blanker. Other types of electric noise defy noise blankers, and the best you can do is hope that a limiter will give some relief by giving the desired signals a "fighting chance" against the noise.

For Nerds Only

In a scheme called *phase cancellation*, noise from a special auxiliary antenna nulls out the noise from the main receiving antenna, in effect making the noise kill itself. The auxiliary antenna actually works best if it *does not* pick up any of the desired signals! I've used that technique at VLF, and it can work amazingly well when everything else fails. The trick lies in making sure that the "noise antenna" picks up lots of noise but little or no signal, and ensuring that its output can be adjusted so that it perfectly "bucks" the noise picked up by the main antenna.

The Worst Noise of All: EMP

An *electromagnetic pulse* (EMP) is a sudden burst of EM energy caused by a single, abrupt change in the speed or position of a group of charged particles. An EMP has no well-defined frequency or wavelength; it exists over the entire EM spectrum including radio wavelengths, infrared, visible light, ultraviolet, X rays, and gamma rays. An EMP can be generated by arcing, and on a radio receiver, multiple small EMPs sound like popping or static bursts.

An EMP can contain a fantastic amount of power for a short time. Lightning discharges have been known to induce current and voltage spikes in nearby electrical conductors, of such magnitude that equipment is destroyed and fires are started. If a solar disruption suddenly sends a huge quantity of charged subatomic particles in the earth's direction, an EMP can occur when the "tsunami" of particles encounters the earth's magnetic field, which accelerates the charge carriers toward the geomagnetic poles.

The detonation of an atomic bomb creates a strong EMP. The explosion of a multimegaton hydrogen bomb at a very high altitude, while not creating a devastating shock wave or heat blast at the surface of the earth, could generate a disruptive EMP over an entire nation or a large part of a nation. The resulting voltages and currents in radio antennas, telephone wires, and power transmission lines could damage or destroy sensitive electronic components connected to them.

Heads Up!

All communications systems should incorporate some form of protection against the effects of an EMP. One possible means of defense is provided by equipment for protection against direct lightning hits. Unfortunately, as of this writing, a lot of media hullabaloo has been stirred up over the danger posed by this phenomenon, but few (if any) countries have undertaken serious preventive measures on a large scale.

CHAPTER 6

Mobile and Portable Ham Stations

When lay people think of Amateur Radio, they usually imagine diverse desktop "rigs" and big antenna "farms." When most folks think of mobile stations, they envision radios of the sort found in taxicabs or police vehicles. The notion of portable radio operation brings to mind the "walkie-talkies" used by security personnel or young children. Amateur Radio offers all those hardware options and more, in the form of mobile and portable equipment.

W1GV Travel Log

In February 2014, I acquired a Yaesu FT-857D mobile/portable transceiver (Fig. 6-1), which covers 160 meters through 70 centimeters and can work in all popular modes. My vehicle already had a 2-meter FM mobile radio, also by Yaesu (Fig. 6-2). I got a substantial mobile antenna base mount, a quarter-wave CB whip chopped down to resonate on 28 MHz, a collapsible whip from MFJ Enterprises that extends from 0.7 meter to 5 meters, an Active Tuning Antenna System (ATAS) from Yaesu, a 35-ampere-hour, deep-cycle marine battery from Batteries Plus, and a keyer paddle from Bencher. Then I took off to Wyoming on a winter vacation, and had mobile and portable CW QSOs with stations all over the world. As this chapter progresses, I'll relate a few of the most notable moments to give you an idea of how much fun ham radio operators can have "on the road" and "in the field"!

Mobile Band Options

For mobile operation, the VHF and UHF bands generally work better, from a technical standpoint, than the HF bands do, especially the ones at 1.8 MHz, 3.5 MHz, and 7 MHz. That's because, as the wavelength gets longer, the minimum physical length or height of an efficient antenna increases, making a series-connected inductance

Figure 6-1   A mobile/portable ham radio transceiver that covers 160 meters through 70 centimeters in all popular modes.

Figure 6-2   A mobile 2-meter FM transceiver.

mandatory. But putting a coil in the radiating element of an antenna reduces its efficiency (the ratio of the RF power that actually goes out into space to the RF power that your transmitter puts in at the feed point). In mobile stations, a marginal ground system, usually comprising only the vehicle body, compounds the problem.

Mobile Antenna Considerations

In practical situations, you can't operate mobile stations below about 24 MHz without placing an inductor in the antenna element. A full quarter-wavelength vertical antenna at 24 MHz measures 3 meters tall, while the same size antenna, in terms of its wavelength dimension, measures 10 meters tall at 7 MHz and 20 meters tall at 3.5 MHz. When the height of a mobile antenna exceeds 3 meters, it becomes unwieldy and unsafe. It will not likely survive the hurricane-force winds that assault it at freeway speeds, and it can strike low-hanging obstructions, such as tree branches, utility wires, small bridges, and tight overpasses. Besides all that, a mobile antenna taller than 3 meters will raise police and troopers' eyebrows, if not provoke them to pull you over and give you a ticket for creating a hazard on the road.

You can get around the practical height limitation for mobile antennas by inserting a coil in the radiating element of a short antenna to lower the resonant frequency, making it electrically longer without making it physically taller. Ideally, the coil should go near the middle of the element, although it can work at any point from the base to roughly three-quarters of the way to the top. When you employ so-called *inductive loading* this way, you lose more power in the ground system than you would lose if you fed a full-size, quarter-wavelength element directly at the base. A midsize car can serve as a decent RF ground at 2 meters or 6 meters, but the efficiency of that same ground system grows worse as the wavelength increases until, at 80 meters or 160 meters, that same vehicle provides almost no RF ground.

In general, if you have a radiating element of a fixed length, you should do everything you can to minimize the *loss resistance*. That means you need a good ground system. Unfortunately, in a mobile station at a wavelength longer than 12 or 15 meters, a good ground system is all but impossible to attain unless you're piloting a big truck, a train, an aircraft, or a large boat. You can't expect to get an efficient HF mobile antenna in most situations with ordinary cars. But at VHF and UHF, you can do quite well! Some vendors offer 1/2-wave or 5/8-wave VHF and UHF mobile antennas that optimize the efficiency by maximizing the so-called *radiation resistance*.

Tip

In Chap. 7, I'll explain how radiation resistance and loss resistance affect the efficiency of an antenna system. For now, keep in mind that you'll do best in terms of antenna efficiency if you use as large an antenna as possible, operate on the highest workable frequency, and drive the biggest vehicle you can find!

Advantages of VHF and UHF Mobile

When you decide to set up a mobile ham radio station, you have a choice between VHF (the bands at 50 MHz and up) or HF (the bands below 29.7 MHz). Here are some good reasons why you might prefer to choose VHF:

- Mobile antennas almost always exhibit good efficiency at VHF and UHF.
- Antennas for VHF and UHF have manageable size.
- You can usually get away with a magnetic mount at the base of a VHF or UHF mobile antenna (for example, the one shown in Fig. 6-3). Such a mount doesn't force you to deface the vehicle, and you can easily and quickly remove it when you want the antenna off the vehicle (for example, in a car wash or garage).
- You'll find plenty of repeater activity on VHF and UHF, especially on 2 meters and 70 centimeters, and these systems offer great convenience as long as you know the tone squelch frequency for every repeater that you want to use.
- Most VHF and UHF radios are simple and compact, don't gobble up a lot of power, and lend themselves to easy operation.
- Ignition noise and power-line noise rarely pose serious problems for reception at VHF and UHF, although vehicle alternators can sometimes produce an annoying whine in the receiver and a similar whine in your transmitted audio.

FIGURE 6-3 A magnetic base mount (also called a "mag mount") for a VHF mobile ham radio antenna.

Advantages of HF Mobile

Some radio hams aren't satisfied with VHF for mobile operation. Here are a few reasons why you might want to install an HF mobile radio in your vehicle instead of, or in addition to, a VHF rig:

- The HF bands, especially the ones at 14 MHz and up, offer fascinating potential for making DX contacts from your vehicle.
- You can almost always contact someone on some HF band from anywhere in the world, no matter how remote your location. This capability can prove invaluable when you drive through places where neither cell phone nor repeater coverage exists.
- Antennas for mobile HF operation, while less efficient than those for VHF operation, can give you plenty of contacts even if they waste considerable RF power in the process.
- The HF bands don't have "closed repeaters" that can render VHF operation useless and pointless for hams who don't know the tone squelch frequency for the repeater nearest them.
- Mobile HF radios can serve as emergency communications stations, especially on the high seas and in aircraft. If you also have VHF and UHF capability, you have the best of everything this respect.

W1GV Travel Log

Surpassing all the above-mentioned advantages of HF mobile operation, I'll add my own personal one: It's fun! I glued a Morse code keyer paddle to a piece of wood and set it on the armrest between the front seats, as shown in Fig. 6-4. Then I worked stations all over the planet on mobile CW from Wyoming on vacation in February of 2014. Using that paddle while driving proved no more distracting or awkward than shifting the gears. But take note: I pulled into parking lots and stopped when I had to do anything that involved fine adjustments to the radio, such as RF power output or keyer speed. I never let my desire for a good time trump my desire to remain healthy and alive!

Warning!   If you plan to operate a mobile ham radio station, check with state and local law enforcement officials before doing it while the vehicle is in motion. You can of course use good judgment, but that will only take you so far these days. As state governments grow more and more eager to legislate common sense, more and more regions and municipalities will likely forbid some types of ham radio mobile operation, if not all of it (except while parked).

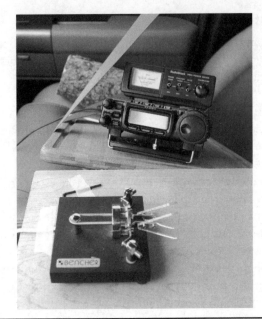

FIGURE 6-4 The W1GV mobile CW installation, showing the radio and SWR meter in the background and the keyer paddle in the foreground.

Mobile Power Options

When you choose a radio for your mobile station, you can get one that runs relatively high power (which hams sometimes call *QRO*, the Q signal for "increase power"), or you can get a radio that runs strictly QRP. For the purposes of this discussion, let's call an RF power output of 25 watts or more QRO, and anything less than 25 watts QRP.

An Old Timer's Opinion

I recommend that you get a QRO-capable radio for your mobile applications. In most cases, you can scale the power back in a QRO transceiver, but you can't increase it to QRO levels if you have a QRP-only radio.

Advantages of QRO Mobile

Most mobile operators, especially when operating in the HF bands, prefer to run the standard RF power of about 100 watts that most transceivers put out. The majority of commercially available VHF radios can run at least 25 watts. Radios get generally less powerful at UHF and microwave frequencies, though. Here are some reasons to run QRO mobile:

- You'll have an easier time making contacts running QRO than you will have if you run QRP.
- With QRO, you'll have a better chance of maintaining a contact once you've made it, especially in light of the fact that mobile operation often involves moving through zones where lines of sight change because of hills and buildings.
- In an emergency, you'll have the best chance of getting help in the least amount of time if you run the highest RF power output that you (and your vehicle's electrical system) can manage.
- Most vehicles' electrical systems can support radios that consume up to about 500 watts, which translates to about 200 watts RF output. The radio-frequency interference (RFI) issue, however, may make it impractical to run that much power, as described below under "Advantages of QRP mobile."

Warning!   Some vehicles' electrical systems have trouble even with radios that consume moderate amounts of power. Beware, especially if you have a small car. If you have doubts about your vehicle's electrical system, consult an auto mechanic or the vehicle manufacturer.

Advantages of QRP Mobile

If you'd rather stick with QRP when operating mobile, you should expect challenges, especially if you operate on the lower HF bands. Nevertheless, QRP mobile has plenty of assets, and for hams who enjoy QRP, the technical challenges constitute a good reason to go this route!

- Your radio places little or no strain on your vehicle's electrical system if you run QRP, even if you transmit in a mode that involves continuous carrier output, such as FM.
- With QRP, your radiated signal will not likely cause the vehicle's computer to malfunction because of RFI. At power levels over 25 watts at some frequencies, strange manifestations can occur, such as speedometer and tachometer reading gyrations on the less serious side, and fuel injection problems or ignition failures on the more serious side.
- You can take your DC power from a cigarette lighter with an adapter plug, such as the one shown in Fig. 6-5, because QRP doesn't demand much current, and you needn't worry about possible contact resistance in the connection, which can prove problematic or even dangerous when you use that type of adapter with a QRO radio.
- You can use a battery other than the vehicle battery to power the transceiver. That way, you minimize the risk of ignition noise and alternator whine coming into your radio through the DC power cable.

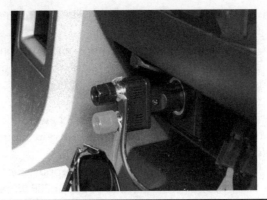

Figure 6-5   A DC power adapter plug for low-current mobile. It's the same type of plug that you find in cell-phone and tablet device chargers.

Tip

You can increase the workable bandwidth of an inductively loaded mobile antenna by placing a transmatch at the transceiver's antenna connection, between the radio and the feed line. You can use a transmatch in the same configuration to "force feed" an antenna and its transmission line on a band whose frequency differs from than that of the band for which the antenna is cut. I used a PalStar antenna tuner to make my 10-meter whip antenna work from 6 meters all the way down to 40 meters (Fig. 6-6). You'll suffer some loss in the feed line when you take advantage of this trick, but it won't amount to much because in mobile applications the feed line is short, and therefore, the line will have low loss even if it's not well matched to the antenna. You must make sure, however, not to run more than about 50 watts RF output if you use small coax for the feed line; otherwise, high currents or voltages can cause heating and arcing that destroy the line's dielectric.

Portable Band Options

In ham radio, the term *portable* can have either of two meanings. Let's call them *heavy portable* and *light portable*. You can easily move the hardware for a heavy portable station from one fixed QTH to another, but you can't use the gear as you carry it. A light portable station, in contrast, lends itself to use as you haul it around. Heavy portable stations can work on HF, VHF, or UHF, but light mobile stations usually work only on VHF and UHF.

Examples

My heavy portable station originally comprised a Yaesu FT-857D compact all-band transceiver along with a deep-cycle marine battery, a keyer paddle glued to a piece

FIGURE 6-6 A transmatch in a mobile installation for increasing antenna bandwidth and operating on bands at frequencies far different than that of the antenna's design band. In my arrangement, the radio sits on top of the transmatch.

of wood to serve as an armrest, and a meter that indicated how well the radio was matched to the antenna (called an *SWR meter*, where *SWR* means *standing-wave ratio*.) as shown in Fig. 6-7. Later I replaced the SWR meter with the transmatch shown in Fig. 6-6, which has its own built-in SWR and power meter. This station doubles as the mobile installation I could also use while camping, or for an emergency situation, or for the annual ARRL Field Day contest, but I couldn't walk around with it and operate it at the same time. I deployed it on the HF bands from the guesthouse at the Stargazer Ranch in Meeteetse, Wyoming, and enjoyed great DXing with 50 watts output on 10, 12, 15, 17, and 20 meters.

A light portable station almost always takes the form of a handheld transceiver. These units look something like the toy "walkie-talkies" that children use to communicate over short distances. But ham radio handheld units or "handy-talkies" incorporate sophisticated features such as microprocessor control and digital capability, which walkie-talkies rarely have. This type of radio has a lightweight battery inside the box. Once in awhile you'll find someone walking around with a more substantial radio, such as the Yaesu FT-857D and a small, light battery in a backpack, but such operations are generally confined to VHF and UHF because of antenna constraints. In the extreme, you might see someone hiking on the trail with a 10-meter whip antenna sticking up out of a backpack! But you'll never see a light portable station in operation with a big, bulky battery such as the marine battery that I use in my heavy portable station.

Figure 6-7   A heavy portable station deployed in the dining area of the bunkhouse at the Stargazer Ranch in Meeteetse, Wyoming, March 2014.

Portable Antenna Considerations

If you plan to do heavy portable operation, you'll need an antenna that you can easily transport and quickly set up at the destination or stopover QTH. A length of coaxial cable with a rolled-up dipole antenna can serve this purpose, but you'll need to take advantage of existing supports such as trees to get anything out of it. Alternatively, you can place a full-size quarter-wavelength vertical antenna on your vehicle and run a length of coaxial cable from your rig to it. Obviously, you must remove or collapse such an antenna while driving the vehicle, but when it's parked, anything goes!

> **Tip**
> A random length of wire (the longer the better) with an antenna tuner at the transmitter offers a great option for heavy portable installations, provided you can get a good ground for the antenna to "work against" and can find something to support its far end a reasonable distance above the earth's surface.

I obtained a collapsible whip antenna from MFJ Enterprises that extends continuously up to 5 meters tall, so I can set it to any frequency on any ham band from 6 meters through 20 meters. It has a standard threaded base of the same sort used with CB whips. I placed the antenna in a so-called *ball mount* secured to the lip of my pickup truck bed, as shown in Fig. 6-8. The ball mount can withstand the stress of gale-force winds that commonly take place in locations I like to frequent, and also facilitates connecting the braid (shield) of the coaxial cable to the vehicle frame. Figure 6-9 shows the whole antenna with the lowermost two sections extended.

The feed line for the antenna shown in Figs. 6-8 and 6-9 comprises a ready-made length of RG-213 coaxial cable (the best kind commonly available for radio hams), 33 meters long with molded connectors on either end. Even though I never parked my vehicle anywhere near 33 meters from the station, I didn't want to cut into that cable for fear of long-term water intrusion, so I coiled up the portion of cable that I didn't need and set it in the truck bed near the antenna. In order to adjust the frequency of the antenna, I used an SWR-and-power meter that I bought at a Radio Shack retail outlet. I attached the meter, shown in Fig. 6-10, to the top of the transceiver with "sticky pads," making sure not to obscure the speaker or any of the ventilation openings.

Advantages of VHF and UHF Portable

For light or "walk-around" portable operation, VHF and UHF predominate. At lower frequencies, efficient antennas are too large to carry and use at the same time. While you can, in theory, inductively load a shortened quarter-wave antenna to take power on any frequency, you can't expect it to work in a light portable

FIGURE 6-8   Base mount for the HF portable whip antenna.

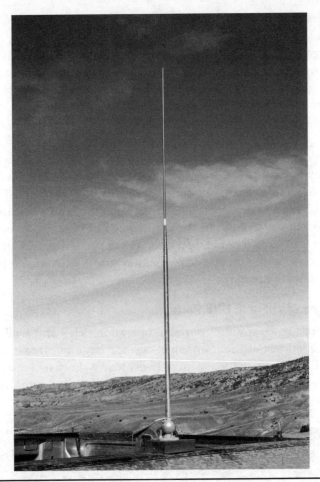

Figure 6-9   The portable whip antenna, almost fully collapsed.

situation because such an antenna needs a substantial RF ground, and an effective "walk-around ground" hasn't been invented yet! Nevertheless, the ham radio spectrum at 6 meters and shorter wavelengths offers specific advantages for any type of portable operation.

- You can make a portable antenna exhibit good efficiency at VHF and UHF, even in the absence of a substantial ground system. The best types are half-wavelength affairs with self-contained tuning networks at the base. Best of all is the so-called *J pole*, which comprises a quarter-wave, parallel-wire line with a half-wave radiator connected to one of the top ends of the line.
- Antennas for VHF and UHF have manageable size (except perhaps for a half-wave or J pole antenna at 6 meters), so you can carry them around without trouble.

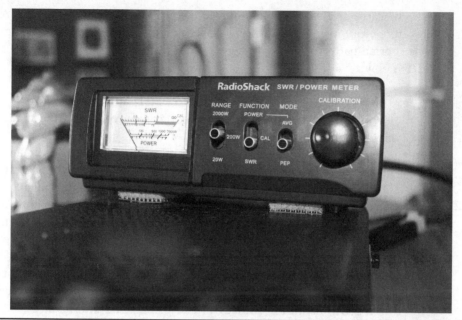

FIGURE 6-10   An SWR-and-power meter facilitates precise adjustment of the portable station antenna frequency and also gives a good indication of actual transmitter output power.

- On the bands at 2 meters and above, the antenna can actually form part of the radio unit, as with a handy-talkie.
- You can put a radio in a backpack for light portable use, and let the antenna stick up out of the backpack for 2 to 3 meters without trouble. The microphone, obviously, has to be used outside the backpack, but that's easy to arrange.
- You'll find repeater activity on 2 meters and 70 centimeters except in the most remote locations. These installations can save your life if you get into serious danger, as long as you stay within range of one or more (and have learned its tone-squelch frequency in advance).
- Most VHF and UHF radios are simple, compact, and easy to operate. At least one radio, the FT-897 made by Yaesu, has a self-contained rechargeable battery and can double as a QRP mobile transceiver. It will also work on HF as a "little sister" to my QRO transceiver, the Yaesu FT-857D.

Advantages of HF Portable

For some hams (myself included), nothing surpasses the thrill of working DX with a portable station on the HF bands. During my week at the Stargazer Ranch in the winter of 2014, I got a great chance to do it! An unusual over-the-pole opening at night on frequencies up to 24 MHz made for fascinating contacts with other hams in such places as Russia, Japan, and South America. I did all this on CW with 50 watts output, using the Yaesu FT-857D powered by a 35-ampere-hour, deep-cycle marine battery. The advantages of HF portable operation resemble those of HF mobile.

- The HF bands, especially the ones at 14 MHz and up, offer great potential for making DX contacts from places far removed from the madness of civilization.
- You can almost always contact someone on some HF band from anywhere in the world. You'll appreciate this fact when you spend a few days at a QTH where cell phones, the Internet, and repeaters aren't available.
- Antennas for portable HF operation can work just as well as those at your home station. In my case, although the antennas were a little less efficient, the location proved superior because humanmade noise hardly existed, and I had a clear shot to the horizon in all directions.
- In many cases, a portable QTH can turn out to be far better than your own home QTH, as mine was at the Stargazer. In particular, you can get away from most, if not all, sources of electromagnetic interference (EMI) that bedevil urban and suburban hams' reception at HF.
- Portable HF radios can serve as emergency communications stations. You can, if necessary, prove your worth in the public service arena during and after a natural disaster, such as a hurricane, flood, wildfire, or earthquake.

An Old Timer's Trick

You can mount a CW keyer paddle on a board so that you can use it on any surface, even in your lap, while working portable. I "cannibalized" a finished shelf from a kitchen cabinet to serve this noble purpose. It measures 600 centimeters long by 250 centimeters wide. It looked and worked great on the dining room table at the Stargazer Ranch guesthouse (Fig. 6-11).

Portable Power Options

As with mobile, you can get a radio that runs either QRO or QRP to work portable. Your choice will depend largely on whether you go for heavy or light operation. You can buy a QRO radio for heavy portable and scale it back for QRP if you want; in light portable the radio will likely not have QRO capability anyhow.

Figure 6-11   A Bencher CW keyer paddle, glued down with contact cement to a wooden board for portable operation.

Advantages of QRO Portable

Some portable operators run the full legal RF power output limit of 1500 watts PEP in heavy portable scenarios, such as the annual ARRL Field Day, in which large gasoline- or propane-fueled generators are common. If you want to use a battery, however, you'll want to run no more than about 100 watts of RF output. Obviously, your battery's life between charging sessions varies in inverse proportion to the current you demand from it. I use a deep-cycle marine battery (Fig. 6-12) with enhanced connections (Fig. 6-13) for the best possible electrical contact, ensuring that problems won't occur at maximum current drain, which in my case can run upwards of 20 amperes with the radio key-down. Here are some reasons to run QRO (over 25 watts output) in heavy portable applications:

- You'll have an easier time making contacts running QRO than you will have if you run QRP, assuming all other factors remain constant.
- With QRO, you'll have a better chance of maintaining a contact once you've made it; you'll be less prone to having QSOs cut short because of interference or fading.

FIGURE 6-12   The deep-cycle marine battery that I use for heavy portable operation. It has a 35-ampere-hour storage capacity and can deliver the 22 amperes necessary to run the Yaesu FT-857D at full RF power output (100 watts).

- In an emergency, you'll have the best chance of getting help in the least amount of time if you run the highest RF power output that you (and your battery, if you use a battery) can manage.

Warning!   When you set up a portable station, make sure that you locate the antenna in such a way that it won't create a hazard. Don't run it over or under electrical power lines. Choose a site that won't make it easy for people to come into contact with the radiating element(s), either accidentally or on purpose. Beware of sudden thundershowers that can expose you and your equipment to danger from induced currents or catastrophic voltages from lightning strikes on or near the antenna. You might want to deploy the antenna in such as way that you can take it down in a hurry should the need arise.

Advantages of QRP Portable

If you'd rather stick with QRP when operating mobile, more power to you (pun intended)! But you should expect challenges, just as you would endure with QRP mobile operation. Here are some of the advantages of going QRP for portable operation.

Figure 6-13 Modified and enhanced battery terminal to facilitate easy insertion and removal of the DC cable wires and to ensure good electrical contact to carry the current that the transceiver demands when "key-down."

- Your radio places minimal strain on a battery if you run QRP, even if you transmit in a mode that involves continuous carrier output, such as FM. The battery will last longer between charges and will be less likely to suffer from a voltage drop because of its own internal resistance.
- With QRP, your radiated signal is not likely to cause RFI to sensitive electronic devices that might exist on the premises (such as heart pacemakers!).
- With QRP, you'll be less likely to have problems with "RF in the shack" if you use an end-fed random wire antenna.
- Most QRP stations are less massive and contain fewer individual boxes than QRO stations, making QRP stations easier to deploy and dismantle.

Stay Safe!

In closing this chapter, please let me reiterate a mandate: Always place your personal safety, and the safety of those around you, above your desire to have "cool fun" with your radio. Don't try to adjust your radio's microcomputer-controlled, menu-driven functions while driving a vehicle. Pull over somewhere, preferably into a parking lot, and make your adjustments there. In heavy portable operations, avoid the temptation to keep operating as a thundershower approaches. In a contest, you might find this point difficult to remember—until sparks start to fly from the back of your radio, or your whole station explodes in your face.

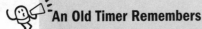
An Old Timer Remembers

When I was a foolish teenage ham in the late 1960s, I loved the ARRL Field Day contest that took place on the last full weekend in June. In southeastern Minnesota where I lived, late June was prime time for thunderstorms and tornadoes. All through my childhood, I marveled at the power of those storms, some of which rivaled tropical cyclones in ferocity. One night, as a thunderstorm swooped across the prairie and slammed into Rochester, Minnesota, I happened to be on 40-meter CW, working stations at a respectable pace. For some time, the fun outweighed my desire to remain alive. Only when sparks began to leap from the back of the antenna tuner— only then!— did someone convince me that I ought to consider going QRT (the Q signal for "Stop transmitting" or "Go off the air") for a while. I obviously had higher powers on my side that night, or you would not have this book in your hands right now. Don't act like an idiot, as I did that night (and more than a few times since)!

FIGURE 6-14   The Active Tuning Antenna System (ATAS) for mobile use with Yaesu radios such as my FT-857D. It covers the bands from 40 meters through 70 centimeters.

FIGURE 6-15   The base mount that I devised for the ATAS. Note the ground wire going from the base of the ATAS to a screw in the truck bed.

Postscript

After returning from Wyoming in the winter of 2014, I installed the Active Tuning Antenna System (ATAS) that I originally got along with the FT-857D. Figure 6-14 shows the general contour of this antenna, which employs a mechanical device to tune the antenna to resonance anywhere from 40 meters (7 MHz) up through 70 centimeters (450 MHz). To support that antenna, I used a Comet mag mount epoxy-glued to a piece of wood, which I then C-clamped to the lip of the truck bed (Fig. 6-15). I connected a ground wire from the shield part of the coaxial cable at the base mount to the truck bed using a short wire with a screw, wing nut, and large washer. After deploying an ohmmeter between various pairs of points to make certain that the ground continuity prevailed throughout the entire truck body, I felt good to go, but as of this writing, had not yet tested the system, except to verify that it will indeed tune on 15 meters and allow me to contact Russia with 25 watts output from a supermarket parking lot! I'll relate the performance of this antenna on my YouTube channel, probably in the "Amateur Radio" playlist. I have uploaded a lot of ham radio videos to that venerable site. I invite you to get on YouTube and subscribe to my channel right now!

CHAPTER 7

Ham Antenna Primer

Ham radio antennas fall into two categories: receiving types and transmitting types. Nearly all transmitting antennas can receive signals effectively within the design frequency range. Some, but not all, receiving antennas can transmit signals with reasonable efficiency.

Radiation Resistance

When RF current flows in an electrical conductor, such as a wire or a length of metal tubing, some EM energy radiates into space. Imagine that you connect a transmitter to an antenna and test the whole system. Then you replace the antenna with a resistance/capacitance (RC) or resistance/inductance (RL) circuit and adjust the component values until the transmitter behaves exactly as it did when connected to the real antenna. For any antenna operating at a specific frequency, there exists a unique resistance R_R, in ohms, for which you can make a transmitter "think" that an RC or RL circuit is, in fact, that antenna. Engineers call R_R the *radiation resistance* of the antenna.

Determining Factors

Suppose that you place a thin, straight, lossless vertical wire over flat, horizontal, perfectly conducting ground with no other objects in the vicinity, and feed the wire with RF energy at the bottom. In this situation, the radiation resistance R_R of the wire is a function of its height in wavelengths. If you graph the function, you'll get Fig. 7-1A.

Now imagine that you string up a thin, straight, lossless wire in free space (such as a vacuum with no other objects anywhere nearby) and feed it with RF energy at the center. In this case, R_R is a function of the overall conductor length in wavelengths. If you graph the function, you'll get Fig. 7-1B.

Antenna Efficiency

You rarely have to think about *antenna efficiency* in a receiving system, but in a transmitting-antenna system, efficiency rises to paramount importance! The

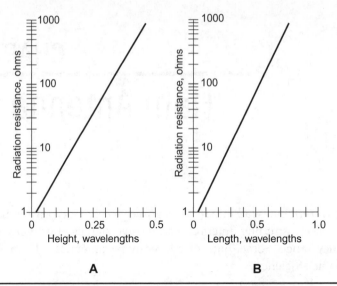

Figure 7-1 Approximate values of radiation resistance for vertical antennas over perfectly conducting ground (**A**) and for center-fed antennas in free space (**B**).

efficiency expresses the extent to which a transmitting antenna converts the applied RF energy to actual radiated EM energy. In an antenna, radiation resistance R_R appears in series with a certain *loss resistance* (R_L), and these resistances behave just like any other types of resistors in series: The larger resistance gets more of the energy than the smaller one does.

In an antenna system, loss resistance "gobbles up" the energy and turns it into useless heat, while radiation resistance puts the energy into the airwaves as your transmitted signal. You can calculate the antenna efficiency, *Eff*, as a ratio with the formula

$$Eff = R_R / (R_R + R_L)$$

If you want to get a percentage figure, the expression becomes

$$Eff_\% = 100\, R_R / (R_R + R_L)$$

You can obtain high efficiency in a transmitting antenna only when the radiation resistance *greatly exceeds* the loss resistance. In that case, most of the applied RF power "goes into" useful EM radiation, and relatively little power gets wasted as heat in the earth and in objects surrounding the antenna.

When the radiation resistance is comparable to, or smaller than, the loss resistance, a transmitting antenna behaves in an inefficient manner. This situation often exists for extremely short antenna radiators because they exhibit low radiation resistance. If you want reasonable efficiency in an antenna with a low R_R value, you must do everything you can to minimize R_L. Even the most concerted efforts rarely reduce R_L to less than a few ohms.

Tip

Even when an antenna system has a high loss resistance, it can still be quite efficient if you design it to exhibit an *extremely high* radiation resistance. When an antenna radiator measures a certain height or length at a given frequency, and if you construct it of low-loss wire or metal tubing, you can get its radiation resistance to exceed 1000 ohms. In that case, you can construct an efficient antenna even in the presence of substantial loss resistance.

Half-Wave Antennas

You can calculate the physical span of an EM field's half wavelength in free space using the formula

$$L_m = 150/f_o$$

where L_m represents the straight-line distance in meters, and f_o represents the frequency in megahertz.

Velocity Factor

The foregoing formula represents a theoretical ideal, assuming infinitely thin conductors that have no resistance at any EM frequency. Obviously, such antenna conductors do not exist in physical reality. The atomic characteristics of wire or metal tubing cause EM fields to travel along a real-world conductor a little more slowly than light propagates through free space. For ordinary wire, you must incorporate a *velocity factor*, v, of 0.95 (95%) to account for this effect. For tubing or large-diameter wire, v can range down to about 0.90 (90%).

Open Dipole

An *open dipole* or *doublet* comprises a half-wavelength radiator fed at the center, as shown in Fig. 7-2A. Each side or "leg" of the antenna measures a quarter wavelength long from the *feed point* (where the transmission line joins the antenna) to the end of the conductor. For a straight wire radiator, the length L_m, in meters, at a design frequency f_o, in megahertz, for a center-fed, half-wavelength dipole is approximately

$$L_m = 143/f_o$$

This formula works based on the assumption that $v = 0.95$, a typical figure for copper wire of reasonable diameter.

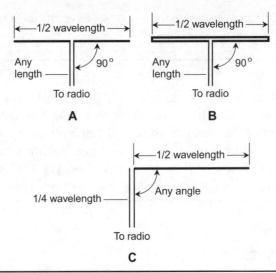

FIGURE 7-2 Basic half-wave antennas. At **A**, the dipole. At **B**, the folded dipole. At **C**, the zepp.

In free space, the impedance at the feed point of a center-fed, half-wave, open dipole constitutes a pure resistance of approximately 73 ohms. This resistance represents R_R alone. No reactance exists, because a half-wavelength open dipole exhibits resonance, just as a tuned circuit would if made with a discrete resistor, inductor, and capacitor.

Tip

When you build a dipole antenna, cut it a couple of percentage points longer than the calculated lengths. Then use a standing-wave-ratio (SWR) meter and a diagonal cutter to trim the antenna, bit by bit from each end, until the lowest SWR occurs at the frequency you most often use (or at the center of the band if you don't have any particular frequency preference).

Folded Dipole

A *folded dipole antenna* consists of a half-wavelength, center-fed antenna constructed of two parallel wires with their ends connected together, as shown in Fig. 7-2B. The feed-point impedance of the folded dipole is a pure resistance of approximately 290 ohms, or four times the feed-point resistance of a half-wave open dipole made from a single wire. This "resistance-multiplication" property makes the folded dipole ideal when using parallel-wire transmission lines.

Half-Wave Vertical

Imagine that you stand a half-wave radiator "on its end" and feed it at the *base* (the bottom end) against an earth ground, coupling the transmission (feed) line to the antenna through an inductance-capacitance (*LC*) circuit called an *antenna tuner* or *transmatch*, designed to cancel out the reactance and transform the remaining resistance to 50 ohms. Then you connect the other end of the feed line to a radio transmitter. This type of antenna works as an efficient radiator even in the presence of considerable loss resistance R_L in the conductors, the surrounding earth, and nearby objects because the radiation resistance R_R is high.

Zepp

A *zeppelin antenna*, also called a *zepp*, comprises a half-wave radiator fed at one end with a quarter-wave section of parallel-wire transmission line, as shown in Fig. 7-2C. The impedance at the feed point is an extremely high, pure resistance. Because of the specific length of the transmission line, the transmitter "sees" a low, pure resistance at the operating frequency. A zeppelin antenna works well at all harmonics (whole-number multiples) of the design frequency. If you use a transmatch to "tune out" reactance, you can use any convenient length of transmission line.

Because of its non-symmetrical geometry, the zepp antenna allows some RF radiation from the feed line as well as from the antenna. That phenomenon sometimes presents a problem in radio transmitting applications. Amateur radio operators have an expression for it: "RF in the shack." You can minimize woes of this sort by carefully cutting the antenna radiator to a half wavelength at the fundamental frequency, and by using the antenna only at (or extremely near) the fundamental frequency or one of its harmonics.

J Pole

You can orient a zepp antenna vertically, and position the feed line so that it lies in the same line as the radiating element. The resulting antenna, called a *J pole*, radiates equally well in all horizontal directions. The J pole offers a low-cost alternative to metal-tubing, vertical antennas at frequencies from approximately 10 MHz up through 300 MHz. In effect, the J-pole is a half-wavelength vertical antenna fed with an *impedance matching section* comprising a quarter-wavelength section of transmission line. The J pole does not require any electrical ground system, a feature that makes it convenient in locations with limited real estate.

Quarter-Wave Verticals

The physical span of a quarter wavelength antenna is related to frequency according to the formula

$$L_m = 75.0v/f_o$$

where L_m represents a quarter wavelength in meters, f_o represents the frequency in megahertz, and v represents the velocity factor. For a typical wire conductor, $v = 0.95$ (95%); for metal tubing, v can range down to approximately 0.90 (90%).

You must operate a quarter-wavelength vertical antenna against a low-loss RF ground if you want reasonable efficiency. The feed-point value of R_R over perfectly conducting ground is approximately 37 ohms, half the radiation resistance of a center-fed half-wave open dipole in free space. This figure represents radiation resistance in the absence of reactance, and provides a reasonable impedance match to most coaxial-cable-type transmission lines.

Ground-Mounted Vertical

The simplest vertical antenna comprises a quarter-wavelength radiator mounted at ground level. You feed the radiator at the base with coaxial cable, connecting the cable's center conductor to the base of the radiator, and the cable's shield to ground.

Unless you install an extensive *ground radial* system with a quarter-wave vertical antenna, it will have poor efficiency unless the earth's surface in the vicinity forms an excellent electrical conductor (salt water, for example). In receiving applications, vertically oriented antennas "pick up" more human-made noise than horizontal antennas do. The EM fields from ground-mounted transmitting antennas are more likely to interfere with nearby electronic devices than are the EM fields from antennas installed high above the ground.

Ground Plane

A *ground-plane antenna* is a vertical radiator, usually 1/4 wavelength tall, operated against a system of 1/4-wavelength conductors called *radials*. The feed point, where the transmission line joins the radiator and the hub of the radial system, is elevated. When you place the feed point at least 1/4 wavelength above the earth, you need only three or four radials to obtain low loss resistance for high efficiency. You extend the radials straight out from the feed point at an angle between 0° (horizontal) and 45° below the horizon. Figure 7-3A illustrates a typical ground-plane antenna.

A ground-plane antenna works best when fed with coaxial cable. The feed-point impedance of a ground-plane antenna having a quarter-wavelength radiator is about 37 ohms if the radials are horizontal; the impedance increases as the radials *droop*, reaching about 50 ohms at a *droop angle* of 45°. You've seen ground-plane antennas if you've spent much time around fixed CB-radio installations that operate near 27 MHz, or if you've done much ham-radio activity in the VHF bands at 50 or 144 MHz.

Coaxial Antenna

You can extend the radials in a ground-plane antenna straight downward, and then merge them into a quarter-wavelength-long cylinder or sleeve concentric with the

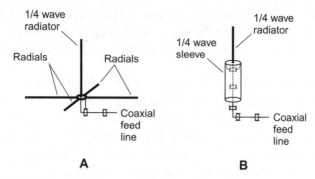

FIGURE 7-3   Basic quarter-wave vertical antennas. At **A**, the ground-plane design. At **B**, the coaxial design.

coaxial-cable transmission line. You run the feed line inside the radial sleeve, feeding the antenna "through the end," as shown in Fig. 7-3B. The feed-point radiation resistance equals approximately 73 ohms, the same as that of a half-wave open dipole. This type of antenna is sometimes called a *coaxial antenna*, a term that arises because of its feed-system geometry, not merely because it's fed with coaxial cable.

Loops

Any receiving or transmitting antenna made up of one or more turns of wire or metal tubing constitutes a *loop antenna*.

Small Loop

A *small loop antenna* has a circumference of less than 0.1 wavelength (for each turn) and can function effectively for receiving RF signals. However, because of its reduced physical size, this type of loop exhibits low radiation resistance, a fact that makes RF transmission inefficient unless the conductors have minimal loss. While small transmitting loops exist, you won't encounter them often because they're hard to design and expensive to build.

Even if a small loop has many turns of wire and contains an overall conductor length equal to a large fraction of a wavelength or more, the radiation resistance of a practical antenna is a function of its *actual circumference* in space, not the total length of its conductors. Therefore, for example, if you wind 100 turns of wire around a plastic hoop that's only 0.1 wavelength in circumference, you'll end up with a low radiation resistance even though the total length of wire equals 10 full wavelengths!

A small loop exhibits the poorest response to signals coming from along its axis, and the best response to signals arriving in the plane perpendicular to its axis. You can connect a variable capacitor in series or parallel with the loop and adjust it until its capacitance, along with the inherent inductance of the loop, produces resonance at the desired receiving frequency. Figure 7-4 shows an example of a tuned small-loop antenna.

Hams sometimes use small loops for *radio direction finding* (RDF) at frequencies up to about 20 MHz, and also for reducing interference caused by human-made noise or strong local signals. A small loop exhibits a sharp, deep *null* along its axis (perpendicular to the plane in which the conductor lies). When you orient the loop just right, so that the null points in the direction of an offending signal or noise source, the unwanted energy will go way down in strength. In some cases the improvement can exceed 20 dB, meaning that the signal power stays the same, while the offending energy gets 100 times weaker!

Loopstick

For receiving at frequencies up to approximately 20 MHz, a *loopstick antenna* can function in place of a small loop. This device consists of a coil wound on a solenoidal (rod-shaped), powdered-iron core. A capacitor, in conjunction with the coil, forms a tuned circuit. A loopstick displays directional characteristics similar to those of the small loop antenna shown in Fig. 7-4. The sensitivity is maximum off the sides of the coil (in the plane perpendicular to the coil axis), and a sharp null occurs off the ends (along the coil axis). Loopsticks are often used along with preamplifiers in high-radio-noise environments to enhance reception. Sometimes, a tiny loopstick can "hear" a distant signal even when a large outdoor antenna can't!

Large Loop

A *large loop antenna* has a circumference of a half wavelength or a full wavelength (depending on the design), forms a circle, hexagon, or square in space, and lies

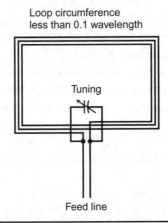

Loop circumference
less than 0.1 wavelength

Tuning

Feed line

Figure 7-4 A small loop antenna with a capacitor for adjusting the resonant frequency.

entirely in a single plane. A well-engineered large loop will work well for transmitting as well as for receiving.

A half-wavelength loop presents a high radiation resistance at the feed point. Maximum radiation/response occurs in the plane of the loop, and a shallow, rather broad null exists along the axis. A full-wavelength loop presents a radiation resistance (and zero reactance, forming a purely resistive impedance) of about 100 ohms at the feed point. The maximum radiation/response occurs along the axis, and minimum radiation/response exists in the plane containing the loop.

The half-wavelength loop exhibits a slight power loss relative to a half-wave open or folded dipole in its *favored directions* (the physical directions in which it offers the best performance). The full-wavelength loop shows a slight gain over a dipole in its favored directions. These properties hold for loops up to several percentage points larger or smaller than exact half-wavelength or full-wavelength circumferences. Resonance can be obtained by means of a transmatch (antenna tuner) at the feed point, even if the loop itself does not exhibit resonance at the frequency of interest.

Here's a Notion!

A *giant loop*, measuring several wavelengths in circumference, can be installed horizontally among multiple supports, such as communications towers, trees, or wooden poles. The gain and directional characteristics of giant loops are hard to predict. If fed with a low-loss feed line using a transmatch at the transmitter end, and if placed at least a quarter wavelength above the earth's surface, such an antenna can offer exceptional performance for transmitting and receiving.

Ground Systems

End-fed, quarter-wavelength antennas require low-loss RF ground systems to perform efficiently. Center-fed, half-wavelength antennas do not, as they, in effect, provide their own RF grounds. However, good grounding, both for RF and for electrical safety, is advisable for any antenna system to minimize interference and hazards.

Electrical versus RF Ground

Electrical grounding constitutes an important consideration if you plan to have good reception and live for a long time to enjoy it! A good electrical (DC and utility AC) ground can help protect your equipment from damage if lightning strikes in the vicinity. A good electrical ground also minimizes the risk of *electromagnetic interference* (EMI) to and from radio equipment.

> **Heads Up!**
>
> In a three-wire electrical utility system, the ground prong on the plug should never be defeated because such modification can result in dangerous voltages appearing on exposed metal surfaces. It can also increase the risk of your suffering with poor radio reception because of humanmade interference sources, such as rogue electrical appliances.

Radio-frequency (RF) *grounding* is a different "beast" from electrical grounding! You'll want to have this asset if you want to avoid "RF in the shack" problems, especially if you have an unbalanced antenna system, such as a ground-mounted vertical or an end-fed, random wire. Figure 7-5 shows a proper RF ground scheme (at A) and an improper one (at B). In a good RF ground system, each device goes to a common *ground bus*, which in turn runs to the earth through a single conductor. This conductor should be as short as you can manage. A poor system has *ground loops* that act like loop antennas, giving rise to all sorts of havoc.

Radials and the Counterpoise

A surface-mounted vertical antenna should employ as many grounded radial conductors as possible, and you should make them as long as you can. You can lay the radials right down on the surface, or bury them a few inches (or centimeters) underground. In general, as the number of radials of a given length increases, the overall efficiency of any vertical antenna improves (if all other factors remain constant). Also, as the radial length increases and all other factors remain constant, vertical-antenna efficiency improves. The radials should converge toward, and connect directly to, a ground rod at the feed point, where the transmission line meets the antenna radiator.

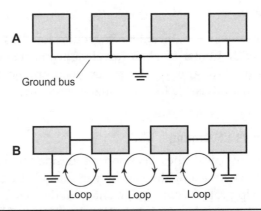

Figure 7-5   At **A**, the correct method for RF grounding of multiple hardware units. At **B**, an incorrect method for grounding multiple equipments, creating RF ground loops.

A specialized conductor network called a *counterpoise* can provide an RF ground without a direct earth-ground connection. A grid of wires, a screen, or a metal sheet is placed above the earth's surface and oriented horizontally to obtain *capacitive coupling* to the earth's conductive mass. The feed point of a vertical antenna goes at the center of the counterpoise. This arrangement minimizes RF ground loss, although the counterpoise won't provide a good electrical ground unless connected to a *ground rod* driven into the earth, or to the utility system ground.

Tip

Ideally, a counterpoise should have a radius of at least a quarter wavelength at the lowest frequency on which you plan to transmit. If you can make the counterpoise bigger than that, then do it!

Gain and Directivity

The *power gain* of a transmitting antenna equals the ratio of the maximum *effective radiated power* (ERP) to the actual RF power applied at the feed point. Engineers express and measure power gain in decibels (dB). Power gain is usually specified in an antenna's favored direction(s).

Power Gain Expressions

Suppose that you symbolize the ERP, in watts, for a given antenna as P_{ERP}, and the applied power, also in watts, as P. Then you can calculate the antenna power gain using the formula

$$\text{Power Gain (dB)} = 10 \log_{10} (P_{ERP}/P)$$

In order to define power gain, you must decide upon some sort of *reference antenna* and define its power gain as 0 dB, meaning no gain or loss at all, in its favored direction(s). A half-wavelength open or folded dipole in free space provides a useful reference antenna for the purpose of comparison with other antennas. Power gain figures taken with respect to a dipole (in its favored directions) are expressed in units called dBd. Some engineers make power-gain measurements relative to a specialized system known as an *isotropic antenna*, which theoretically radiates and receives equally well in all directions in three dimensions, so it has no favored direction. In this case, units of power gain are called dBi.

For any given antenna, the power gains in dBd and dBi differ by approximately 2.15 dB, with the dBi figure turning out larger:

$$\text{Power Gain (dBi)} = \text{Power Gain (dBd)} + 2.15$$

An isotropic antenna exhibits a *loss* of 2.15 dB with respect to a half-wave dipole in its favored directions. This fact becomes apparent if you rewrite the above formula as

$$\text{Power Gain (dBd)} = \text{Power Gain (dBi)} - 2.15$$

Directivity Plots

You can portray antenna *radiation patterns* (for signal transmission) and response patterns (for reception) using graphical plots, such as those in Fig. 7-6. You can assume, in all such plots, that the antenna occupies the center (or *origin*) of a *polar coordinate system*. The greater the radiation or reception capability of the antenna in a certain direction, the farther from the center you should plot the corresponding point.

A dipole antenna, oriented horizontally so that its conductor runs in a north-south direction, has a *horizontal plane* (or *H-plane*) pattern similar to Fig. 7-6A. The *elevation plane* (or *E-plane*) pattern depends on the height of the antenna above *effective ground* at the viewing angle. In most locations, the effective ground comprises an imaginary plane or contoured surface slightly below the actual surface of the earth. With the dipole oriented so that its conductor runs perpendicular to the page or screen (as you look at it in this book or e-reader), and the antenna 1/4 wavelength above effective ground, the E-plane pattern for a half-wave dipole resembles Fig. 7-6B.

Forward Gain

Forward gain is expressed in terms of the ERP in the *main lobe* (favored direction) of a *unidirectional* (one-directional) antenna compared with the ERP from a reference

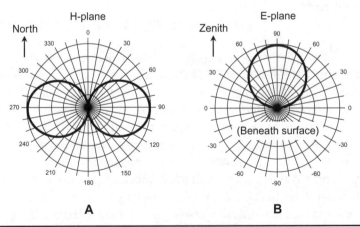

A   **B**

Figure 7-6   Directivity plots for a dipole. At **A**, the H-plane (horizontal plane) plot as viewed from high above the antenna. Coordinate numbers indicate compass direction (azimuth) angles in degrees. At **B**, the E-plane (elevation plane) plot as viewed from a point on the earth's surface far from the antenna. Coordinate numbers indicate elevation angles in degrees above or below the plane of the horizon.

antenna, usually a half-wave dipole, in its favored directions. This gain is calculated and defined in dBd. In general, as the wavelength decreases (the frequency gets higher), you'll find it easier to obtain high forward gain figures.

Front-to-Back Ratio

The *front-to-back* (f/b) *ratio* of a unidirectional antenna quantifies the concentration of radiation/response *in the center* of the main lobe, relative to the direction *opposite the center* of the main lobe.

Figure 7-7 shows a hypothetical directivity plot for a unidirectional antenna pointed north. The outer circle depicts the RF *field strength* in the direction of the center of the main lobe, and represents 0 dB (relative to the main lobe, not a dipole). The next smaller circle represents a field strength 5 dB down (a radiation/response level of −5 dB) with respect to the main lobe. Continuing inward, circles represent 10 dB down (−10 dB), 15 dB down (−15 dB), and 20 dB down (−20 dB). The coordinate origin represents 25 dB down (−25 dB) with respect to the main lobe, and also shows the location of the antenna.

Check It!

If you examine the plot of Fig. 7-7, you can estimate the f/b ratio by comparing the signal levels between north (azimuth 0°) and south (azimuth 180°). That ratio appears to be 15 dB in this case.

Front-to-Side Ratio

The *front-to-side* (f/s) *ratio* provides another useful expression for the directivity of an antenna system. The specification applies to unidirectional antennas, and also to *bidirectional* antennas that have two favored directions, one opposite the other in

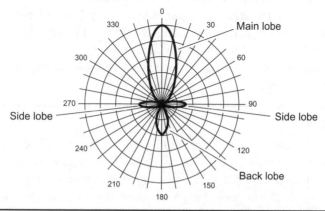

Figure 7-7 Directivity plot for a hypothetical antenna in the H (horizontal) plane. You can determine the front-to-back and front-to-side ratios from such a graph. Coordinate numbers indicate compass-bearing angles in degrees.

space. You express f/s ratios in decibels (dB), just as you do with f/b ratios. You compare the EM field strength *in the favored direction* with the field strength *at right angles* to the favored direction.

Figure 7-7 shows an example. In this situation, you can define two separate f/s ratios: one by comparing the signal level between north and east (the right-hand f/s ratio), and the other by comparing the signal level between north and west (the left-hand f/s ratio). In most directional antenna systems, the right-hand and left-hand f/s ratios theoretically equal each other. However, they sometimes differ in practice because of physical imperfections in the antenna structure, and also because of the effects of conducting objects or an irregular earth surface near the antenna.

Check It Again!

In the situation of Fig. 7-7, both the left-hand and right-hand f/s ratios appear to be roughly 17 dB. You can see this relation if you examine the concentric circles in the plot, noting how far out from the origin the side lobes extend, compared with the main lobe.

Phased Arrays

A *phased antenna array* uses two or more *driven elements* (radiators connected directly to the feed line) to produce power gain in some directions at the expense of other directions.

End-Fire Array

A typical *end-fire array* consists of two parallel half-wave open dipoles fed 90° out of phase and spaced 1/4 wavelength apart, as shown in Fig. 7-8A. This geometry produces a unidirectional radiation pattern. Alternatively, the two elements can be driven in phase and spaced at a separation of one full wavelength, as shown in Fig. 7-8B, producing a bidirectional radiation pattern. When you design a *phasing system* (also called the *phasing harness*) in any antenna array, you must cut the branches of the transmission line to precisely the correct lengths, taking the velocity factor of the line into account.

Heads Up!

Transmission lines have lower velocity factors than single conductors do. The exact value depends on the line type, the construction method used, and the nature of the dielectric material that separates the conductors. If you need to know the velocity factor for a given transmission line, you should check the manufacturer's specifications. In some cases, it can go down as low as about 65%. You'll learn more about transmission lines later in this chapter.

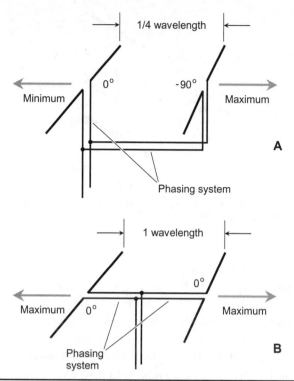

FIGURE 7-8 At **A**, a unidirectional end-fire antenna array. At **B**, a bidirectional end-fire array.

Longwire

A wire antenna, measuring a full wavelength or more and fed at a high-current point (*current loop*) or at one end, constitutes a *longwire antenna*. A longwire antenna offers gain over a half-wave dipole in certain directions. As you increase the length of the wire, making sure that it runs along a straight line for its entire length, the main lobes get more and more nearly in line with the antenna, and their magnitudes increase. The power gain in the main lobes of a straight longwire depends on the overall length of the antenna: the longer the wire, the greater the gain.

You can consider a longwire as a sort of phased array because it contains multiple points along its span where the RF current attains maximum values. Each of these current loops acts like the center of a half-wave antenna, so the whole longwire in effect constitutes a set of two or more half-wave elements placed end-to-end, with each section phased in opposition relative to the adjacent section or sections. As a result, when you have a wire several wavelengths long, you get a complex radiation pattern; as the length grows, so does the complexity!

Broadside Array

Figure 7-9 shows the geometric arrangement of a *broadside array*. Each of the driven elements can comprise a single radiator as shown here, or they can consist of more

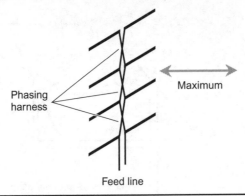

Phasing harness

Maximum

Feed line

Figure 7-9   A broadside array. The elements all receive their portions of the outgoing signal in phase and at equal amplitudes.

complex antennas with directive properties. In this case, all the driven elements are identical, and are spaced at half-wavelength intervals along the phasing harness, which is a parallel-wire transmission line with a half-twist between each element to ensure that all the elements operate in phase coincidence (exactly the same phase).

The directional properties of any broadside array depend on the number of elements (whether or not the elements have gain themselves), the spacing among the elements, and whether or not a reflecting screen is employed. In general, as you increase the number of elements, the forward gain, the f/b ratio, and the f/s ratios all increase.

Tech Tidbit

If you place a flat reflecting screen behind the array of dipoles in Fig. 7-9, you have a so-called *billboard antenna*.

Parasitic Arrays

Communications engineers and radio hams use *parasitic arrays* at frequencies ranging from approximately 5 MHz into the microwave range for obtaining directivity and forward gain. Examples include the *Yagi antenna* and the *quad antenna*. In the context of an antenna array, the term *parasitic* describes the characteristics of certain antenna elements. It has nothing to do with *parasitic oscillation*, a phenomenon that can take place in malfunctioning RF power amplifiers.

Concept

A *parasitic element* is an electrical conductor that forms part of an antenna system but is not connected directly to the feed line. Parasitic elements operate by means of *EM coupling* to the driven element or elements. When power gain occurs in the direction of the parasitic element, you call that element a *director*. When power gain occurs in the direction opposite the parasitic element, you call that element a *reflector*. Directors normally measure a few percentage points shorter than the driven element(s). Reflectors typically measure a few percentage points longer than the driven element(s).

Yagi

A *Yagi antenna*, which radio hams sometimes call a *beam antenna* or simply a *beam*, comprises an array of parallel, straight antenna elements, with at least one element acting in a parasitic capacity. (The term *Yagi* comes from the name of one of the original design engineers.)

You can construct a *two-element Yagi* by placing a director or reflector parallel to, and at a specific distance away from, a single half-wave driven element. The optimum spacing between the elements of a *driven-element/director Yagi* is 0.1 to 0.2 wavelength, with the director tuned 5% to 10% higher than the resonant frequency of the driven element. The optimum spacing between the elements of a *driven-element/reflector Yagi* is 0.15 to 0.2 wavelength, with the reflector tuned 5% to 10% percent lower than the resonant frequency of the driven element. Either of these designs gives you a *two-element Yagi*. The power gain of a well-designed, two-element Yagi in its single favored direction is approximately 5 dBd.

A Yagi with one director and one reflector, along with the driven element, forms a *three-element Yagi*. This design scheme increases the gain and f/b ratio compared with a two-element Yagi. A well-designed, three-element Yagi exhibits about a 7-dBd gain in its favored direction. Figure 7-10 offers a generic example of the relative dimensions of a three-element Yagi. Although this illustration can serve as a crude "drawing-board" engineering blueprint, the optimum dimensions of a three-element Yagi in practice will vary slightly from the figures shown here, because of imperfections in real-world hardware, and also because of conducting objects or terrain irregularities near the system.

The gain, f/b ratio, and f/s ratios of an optimized Yagi all increase as you add elements to the array. You can obtain four-element, five-element, or larger Yagis by placing extra directors in front of a three-element Yagi. When multiple directors exist, each one should be cut slightly shorter than its predecessor.

That's Some Antenna!

Some commercially manufactured Yagis have upwards of a dozen elements. As you can imagine, engineers must spend a lot of time "tweaking" the dimensions of such antennas to optimize their performance. Once you've got a long Yagi tuned just right, you can get well over 10 dBd of forward gain from it.

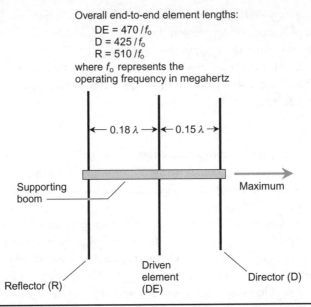

Overall end-to-end element lengths:
$$DE = 470 / f_o$$
$$D = 425 / f_o$$
$$R = 510 / f_o$$
where f_o represents the
operating frequency in megahertz

0.18 λ   0.15 λ

Supporting
boom

Maximum

Reflector (R)   Driven
element
(DE)   Director (D)

Figure 7-10   A three-element Yagi antenna. See text for discussion of specific dimensions. The lowercase, italic Greek lambda (λ) means "wavelength."

Quad

A *quad antenna* operates according to the same principles as the Yagi, except that full-wavelength loops replace the half-wavelength elements.

A *two-element quad* can consist of a driven element and a reflector, or it can have a driven element and a director. A *three-element quad* has one driven element, one director, and one reflector. The director has a perimeter of 0.95 to 0.97 wavelength, the driven element has a perimeter of exactly 1 wavelength, and the reflector has a perimeter of 1.03 to 1.05 wavelengths. These figures represent electrical dimensions (taking the velocity factor of wire or tubing into account), not free-space dimensions.

Additional directors can be added to the basic three-element quad design to form quads having any desired numbers of elements. The gain increases as the number of elements increases. Each succeeding director is slightly shorter than its predecessor. Long quad antennas are practical at frequencies above 100 MHz. At frequencies below approximately 10 MHz, quad antennas become physically large and unwieldy, although some ambitious (and daring) hams have constructed quads to work at frequencies down to 3.5 MHz.

Antennas for UHF and Microwave Frequencies

At frequencies above 300 MHz (wavelengths less than 1 meter), high-gain, multielement antennas have reasonable physical dimensions and mass because the wavelengths are short.

Horn

A *horn antenna* looks like a squared-off trumpet. It provides unidirectional radiation and response, with the favored direction coincident with the opening of the horn. The feed line is a *waveguide* that meets the antenna at the narrowest point (throat) of the horn. (You'll read more about waveguides later in this chapter.)

Horns are sometimes used all by themselves, but they can also feed large *dish antennas* at UHF and microwave frequencies. The horn design optimizes the f/s ratio by minimizing extraneous radiation and response that occur if a dipole is used as the driven element for the dish.

Dish

Most people are familiar with dish antennas because of their widespread use in consumer satellite TV and Internet services. Although the geometry looks simple to the casual observer, a dish antenna must be precisely shaped and aligned if you want it to function as intended. The most efficient dish, especially at the shortest wavelengths, comprises a *paraboloidal reflector*, so named because it's a section of a *paraboloid* (the three-dimensional figure that you get when you rotate a *parabola* around its axis). However, a *spherical reflector*, having the shape of a section of a sphere, will work okay in most dish designs.

A dish-antenna feed system consists of a coaxial line or waveguide from the receiver and/or transmitter along with a horn or helical driven element at the focal point of the reflector. Figure 7-11A shows an example of a *conventional dish feed*. Figure 7-11B shows an alternative scheme known as the *Cassegrain dish feed*. The term *Cassegrain* comes from the resemblance of this antenna design to that of a *Schmidt-Cassegrain reflector telescope*.

As you increase the diameter of the dish reflector in wavelengths, the gain, the f/b ratio, and the f/s ratios all increase, and the width of the main lobe decreases, making the antenna more sharply unidirectional. A dish antenna must measure at least several wavelengths in diameter for proper operation. The reflecting element can consist of sheet metal, a screen, or a wire mesh. If a screen or mesh is used, the spacing between the wires must be a small fraction of a wavelength.

Did You Know?

At microwave frequencies, large dish antennas can have forward gain figures that exceed 35 dBd. That's a power gain of 37 to 1 or more! If you feed such a dish with the maximum legal limit of 1500 watts, you can theoretically get an ERP of 1500 × 37 = 55,000 watts, comparable to a large broadcast station's transmitter output! Don't stand in front of such a monster when it's taking power; the energy coming out of it can cook your insides as if you were the contents of a microwave oven.

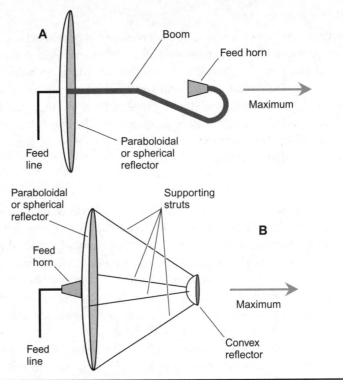

FIGURE 7-11   Dish antennas with conventional feed (**A**) and Cassegrain feed (**B**).

Helical

A *helical antenna* is a high-gain, unidirectional antenna that transmits and receives EM waves with circular polarization. Figure 7-12 illustrates the construction scheme most often used. The reflector must be at least 0.8 wavelength in diameter at the lowest operating frequency. The radius of the helix should be approximately

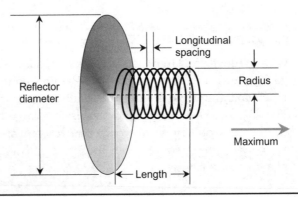

FIGURE 7-12   A helical antenna with a flat reflector.

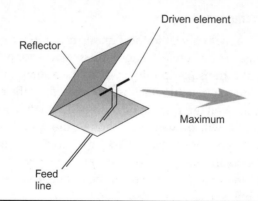

FIGURE 7-13   A corner reflector that employs a dipole antenna as the driven element.

0.17 wavelength at the center of the intended operating-frequency range. The longitudinal spacing between helix turns should be approximately 0.25 wavelength in the center of the operating-frequency range. The entire helix should measure at least a full wavelength from end to end at the lowest operating frequency.

> **Tip**
> When properly designed, a single helical antenna can provide about 15-dBd forward gain. However, they can be phased, just as Yagis or dipoles can, increasing the gain, f/b ratio, and f/s ratio. If you place four helicals at the corners of a square and feed them all in phase, for example, you can get an extra 6 dB of forward gain (a fourfold increase in ERP works out to 6 dB).

Corner Reflector

Figure 7-13 illustrates a *corner reflector* with a half-wave, open-dipole driven element. This design provides some power gain over a half-wave dipole. The reflector is made of wire mesh, screen, or sheet metal. The *flare angle* of the reflecting element equals approximately 90°. Corner reflectors are widely used in terrestrial communications at UHF and microwave frequencies. For additional gain, several half-wave dipoles can be fed in phase and placed along a common axis with a single, elongated reflector, forming a *collinear corner-reflector array*.

Transmission Lines

A *transmission line*, also known as a *feed line*, is a medium through which energy gets transferred from one place to another. In ham radio, the expression refers to the line or cable that connects your transmitter and receiver to your antenna.

Characteristic Impedance

The ratio of the signal voltage to the signal current in a transmission line depends on several things. If you connect a noninductive resistor of the correct ohmic value at the far end of a transmission line, where the antenna would normally go, and then you transmit a signal into the line, not exceeding the power dissipation rating of the resistor, the voltage-to-current ratio will remain constant all along the line. This constant ratio will prevail even though the actual voltage and current will decline slightly as you move away from the transmitter (because of line loss, which every transmission line has). The voltage-to-current ratio under these circumstances is called the *characteristic impedance* or *surge impedance* of the transmission line. Engineers symbolize it as Z_o.

Factoid

The characteristic impedance of a given transmission line depends on the diameter and spacing of the conductors used, and also on the type of dielectric material that separates the conductors.

Coaxial Cable

Coaxial cable is designed for RF signals at frequencies from VLF to UHF. It keeps the desired signals inside itself, and keeps out unwanted signals and interference. Coaxial cable or *coax* (pronounced "CO-ax"), is used by amateur and CB radio operators to connect transceivers, transmitters, and receivers to antennas. It can also be used in high-fidelity audio systems and computer installations to interconnect components. All coaxial cables are fabricated in the same way: A wire is surrounded by a tube or braid of solid or braided metal. The inner wire is called the *center conductor*, and the outer conductor is called the *shield*.

In some coaxial cables, a solid or foamed polyethylene layer, called the *dielectric*, keeps the center conductor running right down the central axis of the cable, as shown in Fig. 7-14A. The dielectric also keeps the center conductor from shorting out to the shield. Other cables have an uninsulated center conductor and a thin layer of polyethylene just inside the braid (Fig. 7-14B), so that most of the interior of the cable comprises air. In this cable, the center conductor can move around somewhat, but the polyethylene keeps it from shorting out to the shield.

Did You Know?

Most coaxial cables use braided copper for the shield, but some cables have a solid metal pipe surrounding the center conductor. This type of cable is called *hardline*. It's expensive, but most hardlines can handle more power, and have lower loss per unit length, than ordinary coaxial cable.

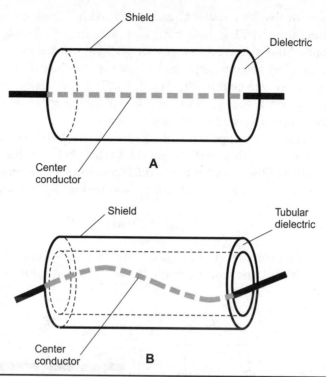

FIGURE 7-14 Coaxial cable with a solid or foamed dielectric (**A**), and coaxial cable with a dielectric comprising mostly air (**B**).

In a coaxial cable, the signal is carried by the center conductor. The shield is connected to an electrical ground. Grounding is important for a coaxial cable to function properly. When the shield of the cable is well grounded, it keeps the data signals from "leaking" out of the cable. That's one of the big advantages of coax. Another advantage is the fact that the shield, when properly grounded, keeps unwanted signals or noise from getting in from sources right next to the cable. Twisted-wire or open-wire lines can let stray RF fields, impulse noise, and appliance noise in, degrading reception.

> **Tip**
> You can run a coaxial cable right next to metal objects, and through areas where there is a lot of interference, and it will still work well. You can't do that with unshielded cables.

Parallel-Wire Line

Parallel-wire line is commonly used with balanced (symmetrical) antennas, such as dipoles and most Yagis. The simplest parallel-wire transmission line is *open wire*. It consists of two wires parallel to each other, with the fewest possible number of intervening supports. The dielectric is almost entirely air. This type of transmission line is one of the oldest, and most effective, ways of getting power from a radio transmitter to an antenna, especially at frequencies below 30 MHz. It's still used today by many Amateur Radio operators. You can make open-wire line from common electrical wire, such as American Wire Gauge (AWG) No. 12 or No. 14 soft-drawn copper, spaced several inches apart.

The characteristic impedance of an air-dielectric balanced line depends on the radius of the conductors, and also on the spacing between the conductors. If you call the radius of the conductors r and the center-to-center spacing d, as shown in Fig. 7-15A, then you can calculate the approximate characteristic impedance as

$$Z_o = 276 \log_{10} (d/r)$$

as long as you express r and d in the same units, such as centimeters or millimeters. The presence of plastic spacers, employed to keep the conductors evenly separated,

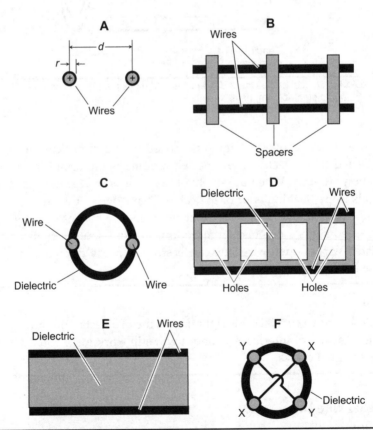

FIGURE 7-15   Open-wire line (**A**), ladder line (**B**), tubular line (**C**), window line (**D**), ribbon line (**E**), and four-wire line (**F**).

will lower this value slightly. Also, if you place the line near objects with a higher dielectric constant than air, the characteristic impedance will go down a little.

Figure 7-15B shows a typical ladder-type, open-wire line. In prefabricated ladder lines, the wires are usually separated by a centimeter or two, and consist of AWG No. 16 or No. 18 soft-drawn uninsulated copper wire. The spacers are placed at regular intervals of about 15 centimeters to keep the wires at constant separation. The characteristic impedance ranges from 300 ohms for a close-spaced ladder line to 450 ohms for the wide-spaced type.

Ladder line has low loss and high power-handling capacity, even in the presence of significant mismatches between the line and the antenna. You'll usually need an antenna tuner (transmatch) having a balanced output at the station end of the line in shortwave and Amateur Radio installations.

Here's a Plug!

As of this writing, a company called *Palstar* makes an excellent antenna tuner designed with a true balanced output. Most balanced-line transmatches use special transformers called *baluns* (a contraction of the expression "balanced-to-unbalanced) to interface coax with two-wire line. That's an engineering compromise. Palstar does it the right way, with duplicate tuned circuits, one for each conductor in the balanced line. Check them out at

www.palstar.com

One of the Palstar transmatches got a favorable review in *QST* Magazine, the official ARRL publication for hams. In my opinion, if the ARRL says that a product is good, then it's got to be good!

Ladder line works well for residential TV reception using old-fashioned outdoor antennas in fringe areas because this line has lower loss per unit of length, than solid-dielectric lines. Ladder line is also relatively immune to changes in characteristic impedance that can occur when it rains. The main trouble with ladder line is that, if it gets twisted, the conductor spacing will change, and the wires can short out because they lack insulation.

Some of the problems with ladder line and open wire are overcome by so-called *tubular line*. This type of parallel-wire line can offer performance comparable to ladder line when properly installed and maintained. Figure 7-15C shows an end-on, cross-sectional view of a tubular line. The conductors are molded diametrically opposite each other in a long, flexible cylinder of polyethylene. Tubular lines have Z_o values similar to those of ladder lines. Most of the effective dielectric is air inside the cylindrical supporting structure. This design attribute minimizes line loss per unit of length, and reduces the effects of wetting or icing on the Z_o. To prevent water from collecting inside the tube, you can punch a couple of small holes in the cylinder at physical low points, so that any water that might collect inside the line can run out. One problem with tubular line is the fact that worms, insects, and spiders can get inside the cylinder, lowering the Z_o especially if the creatures spin webs, which hold condensation. This "bugaboo" (pun intended) also lowers the power-handling capacity and increases the line loss. Some tubular lines are hermetically sealed and filled with inert gas to eliminate this trouble, but such lines are expensive, and they cannot be cut if you want to trim the line length.

Parallel-wire line is commonly molded in the edges of polyethylene strips. Some types are suitable for use in Amateur Radio and shortwave-listening stations. This type of line is called *twinlead* or *ribbon*. Most twinlead lines have a characteristic impedance of 300 ohms when dry, and when not run near conducting obstructions. You can find this stuff in a variety of forms. Figure 7-15D shows "window" type twinlead. Much of the dielectric is cut away to minimize the line loss, while maintaining constant conductor separation. The "windows" also reduce the extent to which rain affects the characteristic impedance of the line. Figure 7-15E shows twinlead with a solid web of dielectric material. The conductors are stranded copper, usually of size AWG No. 14 to No. 20. Some solid-web types of twinlead have a foamed-polyethylene dielectric for reduced loss. Some hams use TV ribbon in low-power transmitting applications at high frequencies (below 30 MHz) in place of open-wire or ladder line. However, the loss per unit of length of such line exceeds that of open-wire or ladder line. This is attributable to the higher dielectric loss of polyethylene compared with air. Solid-dielectric twinlead is affected considerably by wetness and icing, both of which lower the Z_o and increase the loss even more. In addition, receiving type TV ribbon can't handle as much power as a good open wire or ladder line can.

Four-wire line is not used as often as other types of parallel-wire line, although it has significant assets. Four conductors are run parallel to each other, in such a way that they form the edges of a long square prism (Fig. 7-15F). Each pair of diagonally opposite conductors is connected together at either end of the line. You can substitute four-wire line for two-wire line in almost any installation. For a given conductor size and spacing, four-wire line has a lower characteristic impedance than two-wire line; the four-wire line is in fact two two-wire lines connected in parallel. It is difficult to construct a two-wire line with a characteristic impedance of less than 150 ohms, but four-wire line can be made to have a Z_o value as low as 75 ohms. Four-wire suffers less than two-wire line from the effects of nearby objects because more of the RF field stays within the line.

Tip

Four-wire line has inherently better balance than two-wire line does in practical systems, and it's less likely to radiate signals or pick up unwanted noise. However, four-wire line is more difficult to install, is harder to find commercially made, and costs more than two-wire line.

Waveguide

A *waveguide* is an RF transmission line used at UHF and microwave frequencies. In its simplest form, a waveguide consists of a hollow metal pipe having a rectangular or circular cross section. The RF field travels down the pipe, provided that the wavelength is short enough. In order to efficiently propagate RF energy, a rectangular waveguide must have sides measuring at least 0.5 wavelength, and preferably more than 0.7 wavelength. A circular waveguide should measure at least 0.6 wavelength in diameter, and preferably 0.7 wavelength or more. If all of the RF field's electric lines of flux are perpendicular to the axis of the waveguide, the waveguide is said to operate in the *transverse-electric* (TE) *mode*. If all of the magnetic lines of flux are perpendicular to the axis, the waveguide is operating in the *transverse-magnetic* (TM) *mode*. In a waveguide, the characteristic impedance varies considerably with the frequency, unlike other types of line in which the Z_o remains independent of the frequency over a wide range. If you plan to use a waveguide, you must keep its interior clean and free from dust or condensation.

Standing Waves

All transmission lines have some loss; they do not transfer energy with perfect efficiency. The loss occurs because of the ohmic resistance of the conductors, because of *skin effect* (a tendency for RF current to flow mostly near the surface of a conductor), and because of losses in the dielectric. Engineers express this loss in decibels per unit of length. In most lines, the loss per unit of length increases as the frequency goes up. The loss also increases if the load impedance does not match the characteristic impedance of the line.

Standing waves are voltage and current variations that exist on an RF transmission line when the load impedance differs from the Z_o of the line. In a transmission line terminated in a pure resistance having a value equal to the line Z_o, no standing waves occur, and that's the optimal state of affairs. When standing waves do exist, you'll get a nonuniform distribution of RF current and RF voltage as you move along the line. As the impedance mismatch between the line and the load (antenna) gets worse, the voltage and current fluctuations, and therefore, the standing waves, grow more pronounced. The ratio of the maximum RF voltage to the minimum RF voltage, or the maximum current to the minimum current, is called the *standing-wave ratio* (SWR) on the line. The SWR can equal 1:1 only when the current and voltage stay in the same proportions everywhere along the line.

As the mismatch between the line and the load gets worse, the SWR goes up. In theory, there's no limit to how large the SWR on a transmission line can get. In the extreme cases of a short circuit, open circuit, pure inductance, or pure capacitance at the load end of the line, the SWR is theoretically infinite. In practice, line losses prevent the SWR from becoming truly infinite, but it can reach values of 20:1 or more in a real-world transmission line when a short or open circuit, or a pure reactance (capacitance or inductance), takes the place of the load.

Tech Note

In order to get a "perfect match" (1:1 SWR) in your antenna system, you must achieve two goals. First, the load must contain no reactance; it must comprise a pure resistance. Second, the load resistance must equal the Z_o of the line.

The SWR figure can serve as an indicator of the overall performance of an antenna system because a high SWR indicates a severe mismatch between the antenna and the transmission line. This state of affairs can have an adverse effect on the behavior of a transmitter or receiver connected to the antenna system. An extremely large SWR can also cause significant signal loss in a transmission line. If you connect a high-power transmitter to a transmission line with an extremely high SWR, the currents and voltages can get large enough in some regions of the line to cause physical damage to it. The current can heat the conductors to the point that the polyethylene dielectric material melts; the voltage can cause arcing that melts or burns the dielectric.

Tip

Although SWR serves as a significant indicator of antenna design quality, it's not the "end all" figure for analyzing or expressing your antenna's performance. An open-wire line, especially if it's made with large-diameter conductors and has mostly air as the dielectric, can work efficiently even if the SWR is 10:1 or even 15:1! As a matter of fact, if you can get a 2:1 or better SWR at the feed point where the line connects to the antenna, you'll experience a power loss figure of less than 1 dB caused by the mismatch. That amount of *feed line mismatch loss* is so low that you can disregard it "for all intents and purposes," even with a long run of lossy line. So strive for a low SWR, but don't obsess over it.

Stay Safe!

All engineers who work with antenna systems, particularly large arrays or antennas involving the use of long lengths of wire, place themselves in physical peril. By observing some simple safety guidelines, you can minimize (but not completely eliminate) the risk. A few basic tips follow.

- Never manipulate or install an antenna in such a way that it can fall or blow down on utility power lines. Also, it should not be possible for power lines to fall or blow down on an antenna, even in the event of a violent storm.

- Never use wireless equipment having outdoor antennas during thundershowers, or when lightning exists anywhere in the vicinity. Antenna construction and maintenance should never be undertaken if you can see lightning or hear thunder, even if a storm appears far away.
- Disconnect all outdoor antennas from electronic equipment, and connect those antenna feed lines to a substantial earth ground, at all times when you aren't using your station.
- Tower and antenna climbing constitutes a job for professionals only. No inexperienced person should ever attempt to climb any antenna support structure. That includes you, most likely. (It certainly includes me!)
- Indoor transmitting antennas can expose operating personnel to RF field energy. The extent of the hazard, if any, posed by such exposure has not been firmly established. However, sufficient concern exists among some experts to warrant checking the latest publications on the topic. The ARRL can provide you with more information.

Warning!   For complete information on antenna safety matters, you should consult a professional antenna engineer or a comprehensive reference devoted to antenna design and construction. You should also read and heed all electrical and building codes for your city, state, or province. Not only will these precautions minimize physical danger to your person and property, but also it will ensure that your "antenna ranch" doesn't effectively void your homeowner's insurance policy!

CHAPTER 8

Ham Operating Basics

If you're reading this chapter, I assume that you have a ham license, perhaps a new one. If you've just joined our "fraternity," congratulations on getting your license, and welcome to the fray. If you haven't yet bought or built a radio, doubtless you'll do it soon, like later today! In this chapter, after a half century as a participant in the Amateur Radio art, I'll offer a bit of advice concerning how to get the most out of all that new hardware when you get on the air.

Tip

This chapter summarizes ham radio operating basics for most modes and interests, but it only scratches the surface. The best comprehensive reference that I know of for the subject, *The ARRL Operating Manual*, can serve your needs in more detail, once you've read this chapter and decided what you like best. Go to the ARRL website at

www.arrl.org

and look for the latest edition of that fine book. The ARRL has plenty of other publications for specialized ham radio modes and interests, too.

Listen, Listen, Listen!

Has anyone ever told you to have big ears and a small mouth, or something like that? When I was a kid, one of my dad's favorite sayings was "Silence is golden." Well, sometimes that's true, but if nobody ever sent out any squeaks or squawks on the ham bands, we'd lose our privileges for lack of use. Nevertheless, good operators spend more time receiving than they spend transmitting. So listen with abandon, and transmit with care.

Choose Your Band

When you power up your radio, you'll have to choose a band for operation. On VHF, the time of day rarely makes a difference, but on HF it matters a lot. The "upper HF bands" at 20, 17, 15, 12, and 10 meters usually offer better conditions during the daylight hours and in the spring and summer; the "lower HF bands" at 40, 60, 80, and 160 meters usually work better at night and during the fall and winter. (Exceptions happen, of course!) The 30-meter band can go either way, a characteristic that makes that 50-kHz-wide slot especially interesting.

Do you want to have a slow, easy, casual QSO (contact) with someone? Are you planning to work in a contest, or to work DX? Are you involved in an emergency where your operation will take place on a predetermined frequency or frequencies? Do you want to test a new antenna on the band for which it's designed (or maybe on a band for which it's not designed, out of sheer curiosity)? Do you want to use CW, RTTY, DigiPan, SSB, FM, or one of those new and exotic modes, such as MFSK or WSJT? All of these factors will affect which band you choose, and where in that band you operate.

If you have a multiband antenna, you'll hear abundant signals on some bands and few or none on other bands at any given time. But the "spectrumscape" will constantly change!

On a winter night in South Dakota, you'll almost certainly hear signals on 160, 80, and 40 meters. If anyone is using 60 meters you'll probably hear signals there too. But go up to 10, 12, or even 15 meters, and you should expect to hear nothing other than receiver hiss and, if you live in a place where radio-noisy electrical appliances abound (as I do), various forms of human-made roaring, buzzing, and popping.

On a fair afternoon in June, and particularly if the sunspot cycle is near its peak, you should expect to hear plenty of signals on 10, 12, 15, 17, and 20 meters, a few on 40 and 30 meters, very few if any on 60 meters, and only locals and regionals on 80 meters. The 160-meter band might host a few specialized traffic nets, but otherwise, you won't find that band of much use on summer days.

The Dead Band Delusion

The "upper HF bands," especially 10 and 12 meters, occasionally trick us into thinking they're "dead" when in fact they're merely neglected. I've encountered this condition on winter afternoons and evenings. If you spend a couple of minutes tuning through, say, the 10-meter CW band and you don't hear anything, you shouldn't conclude that the band is "dead" until you send a few CQs to see if anyone comes back to you, and have tuned around and listened for a good quarter of an hour. Alternatively, you can tune down to the 27-MHz Citizens Band to see if you hear anything there. If the 10-to-11-meter zone has any life at all, there'll be plenty of "good old boys" having fun in the range of frequencies that I call the "Chaos Band." If you hear signals around 27 MHz, you can have reasonable confidence that the 10-meter ham band is open, whether or not anyone is taking advantage of it! You can, with some patience, have some cool experiences on that "magic band."

The Contest Conundrum

If you operate on weekends, you'll often have to deal with contests that will affect your choice of band, frequency, and/or mode. For example, on specific weekends in November, the Sweepstakes contest fills either the digital or voice parts of most HF bands with aggressive operators using high-powered stations and huge antennas. During the daytime, you might find casual communicating or DXing nigh impossible on 20, 15, and perhaps even 10 meters. At night, you'll encounter the same situation on 80 and 40 meters.

If you want to do some casual operating or DXing or anything else that does not involve contests, and you happen to hit one of the major contest weekends, you can go to a band where contesting doesn't happen. The 60-, 30-, 17-, and 12-meter bands rarely, if ever, host contest events. On the first full weekend in December, you'll encounter the 160-meter contest, and in January, there's a DX contest down there. Otherwise, most weekends will work out all right on that band, at least in the wintertime and at night. And don't forget that you can always do some hamming on weekdays, too!

A Techie's Confession

I love to build and test antenna systems, and I've found contests to be a good way to learn how well they work, especially if I run low power (QRP). If I can "crack a pileup" with 10 watts, I know that my antenna is doing a good job! However, before I call anybody, I always listen to see what the contest exchanges involve, and also, if possible, to find out which contest it is and where it's happening. That way, I won't frustrate serious contest operators who'll want to get me out of the way and put me in their logs with a minimum of "muss and fuss."

Don't Be a Lid!

When a band gets congested, you'll eventually come up against somebody who'll transmit on top of you, insult you, or otherwise act like a jerk. Hams have a term for operators like that: *lid*. Characteristics of a lid, some of which are merely rude and others of which actually violate FCC regulations, include:

- Failing to ask if a frequency is in use before sending a CQ.
- Sending a CQ despite other hams' telling you that the frequency is in use.
- Unidentified transmissions of any sort.
- Sloppy sending on CW.
- Profanity or insulting language in any mode.

- Prolonged, incessant "tuning up" or testing on a single frequency.
- Drawn-out, monologue-like CQs.
- Changing the transmitter frequency while putting out a signal.
- Transmitting a signal whose bandwidth exceeds the legal limit for the mode.
- Deliberately interfering with other hams' QSOs.
- Failing to heed instructions for calling a station when its operator gives specific instructions for calling.
- Failing to yield a frequency to an emergency operation.
- Acting as if you "own" a particular frequency.

No matter how much of a lid someone might be, don't let his or her bad manners turn you into a lid, too. You might feel a strong temptation to transmit "LID" on CW or say, "Get lost, lid" on the phone. Resist that temptation. Only lids call other hams lids on the air.

Signal Reporting

Whenever you make a contact, you'll want to know how strong your signal is at the other end. You'll want to know how you sound, or whether you're causing splatter or otherwise emitting a signal that's not state-of-the-art. This curiosity constitutes part of a good operator's mentality! You'll also want to tell the operator on the other end how well his or her station is doing at your end.

Signal strength reports can vary from qualitative expressions, such as "You're booming in here, the strongest signal I've heard all day!" to quantitative reports that give numbers for readability, strength, and quality of sound or tone. From worst to best:

- Readability numbers go from 1 to 5.
- Strength numbers go from 1 to 9.
- Tone numbers (in CW only) go from 1 to 9.

Table 8-1 breaks down the *RST* (readability, strength, tone) signal reporting system that ham radio operators use. In voice modes, you can leave the tone number out, even though, arguably, it would prove more useful on those modes than on CW! So you might say, in CW, "RST 579" but in SSB you would say "You're 5 by 7."

Despite the lack of a tone number in signal reports for the voice modes, you should tell the other operator if his or her voice is distorted, tinny, muffled, or otherwise mutilated. You can use plain language to describe technical things when working in voice modes. For example, if someone says that your mobile FM signal has *alternator whine*, it means that your vehicle's alternator is causing trouble with your transmitter (modulating your signal with an audible tone that varies in pitch as your alternator speeds up and slows down). If someone says that your FM signal is *full quieting*, it means that you're strong enough to keep the squelch in the receiver completely open so that your signal overcomes the receiver hiss of a partially open squelch.

In CW, the tone number is almost always 9, meaning a pure, steady signal without hum or other modulation that would result from a bad power supply or some sort of unwanted oscillation in one of the transmitter amplifier stages. In fact, the tone number has devolved in the past several decades to the point of irrelevance. Only a primitive transmitter ever puts out a CW tone with quality anything less than 9. If you ever get a tone report of anything other than 9, you had better listen to your signal on a separate receiver, and if a problem exists, get it corrected!

Although CW signals almost always have perfect tone these days, other technical troubles can occur in that mode. Key clicks can mess up a CW transmitter's signal big time. If a station's transmitter generates such artifacts so that you hear popping noises above and below the carrier frequency, add a K at the end of the signal report, for example "RST 579K." Key clicks result from too-fast attack (start) and/or decay (end) periods on code elements. In effect, they amount to CW splatter, and they cause the signal to violate FCC regulations.

TABLE 8-1 The RST (Readability, Strength, Tone) System for Signal Reporting

Number	Readability	Strength *	Tone (CW only) **
1	Utterly unreadable	Barely perceptible	AC modulated with no filtering
2	Barely readable	Very weak	Lots of ripple, only a bit of filtering
3	Readable with difficulty	Weak, poor	Modulated with pulsating DC
4	Almost perfectly readable	Fairly weak	Modulated with severe ripple
5	100-percent readable	Fairly decent	Modulated with strong ripple
6	—	Decent but not spectacular	Modulated with some ripple
7	—	Moderately strong	Modulated with a little ripple
8	—	Good and strong	Modulated with a trace of ripple
9	—	Extremely strong	Pure sine-wave tone, no ripple

*These descriptions represent qualitative notions. Good, honest reports use subjective standards like this, and not the readings from receiver "S" meters. Unfortunately, most signal reports are grossly inflated.

**In recent decades, with the improvements in power-supply technology, any CW tone report other than 9 constitutes cause for embarrassment.

If a CW transmitter produces a carrier that changes frequency slightly at the start of each code element, add a C for "chirp," for example "RST 579C." If a signal is amplitude modulated by an audio tone like a musical note, you'll have to use plain language to inform the other operator. Chirp and musical modulation don't occur often in latter-day equipment, but I've heard both problems on the air as recently as 2013. Usually they appear on the signals from DX stations in countries whose people lack good access to modern technologies.

Did You Know?

In the original RST system, which remained in use until about 1970, one of the tone numbers stood for "musically modulated note." Also, the other tone numbers conveyed meanings more diverse than hair-splitting severity indicators for AC ripple.

Tip

When sending signal reports on CW, an operator will often truncate a number 9 to a letter N. So if you hear me send "5NN SD" the next time you contact me in a contest, it means "599 South Dakota." Contesters use that tactic all the time, and DXers do it a lot, as well. In addition, rather than bothering with a true signal report, contesters and DXers will send "5NN" no matter what the signal sounds like. In such fast-paced operating environments, simply making the contact is the only thing that matters, anyway!

Operating in SSB

On the HF bands, single sideband (SSB) is the most common voice mode. You'll hear mostly lower sideband (LSB) on 160, 80, 60, and 40 meters, while upper sideband (USB) prevails on 20, 17, 15, 12, and 10 meters. You won't, or shouldn't, hear SSB at all on 30 meters.

The Phonetic Alphabet

Voice-mode operation offers the convenience of plain-language communication, so it's more "up close and personal" than digital modes. You can express things without resorting to abbreviations or special signal codes to save time. But once in a while that convenience can turn into a problem, especially under marginal conditions, such as QRM (interference from other stations), QRN (radio noise such as sferics), and QSB (signal fading).

Occasionally, you'll find yourself spelling out certain words because conditions have grown so bad that the other operator has trouble "copying" you on SSB. For example, if someone can't get my name straight (Stan), thinking maybe it's Dan or Sam, I can spell it out for them. But if I say "S-T-A-N," they might instead hear "F-E-A-M." Then they'll be more confused than ever! In situations of this sort, I can use phonetics and say, "My name is Stan: Sierra, Tango, Alpha, November. Stan." That almost always clears things up.

Despite its advantages, some hams use the phonetic alphabet to an extreme. If the other operator has no trouble "copying" you, then you shouldn't use phonetics. And once in a while someone will use phonetics in a way that makes them sound almost lid-like: "CQ Delta X-ray," for example, rather than the correct "CQ DX," which means "Calling any DX station who wants to have a QSO with me."

Tip

You'll find a list of phonetic representations for all 26 letters of the English alphabet in Table 8-2.

Calling a Station

Imagine that you hear someone calling CQ on SSB, or ending a contact while leaving open the possibility of having another one. Wait until the operator stops transmitting, and then push your microphone button and send his or her call once, followed by "this is" and then your own call twice—without phonetics the first time and with phonetics the second time. For example:

W7***, this is W1GV. Whiskey one golf victor. How copy?

TABLE 8-2 Phonetic Representations for Letters of the English Alphabet

Character	Symbol	Character	Symbol
A	Alfa or Alpha	N	November
B	Bravo	O	Oscar
C	Charlie	P	Papa
D	Delta	Q	Quebec
E	Echo	R	Romeo
F	Foxtrot	S	Sierra
G	Golf	T	Tango
H	Hotel	U	Uniform
I	India	V	Victor
J	Juliet	W	Whiskey
K	Kilo	X	X-ray
L	Lima	Y	Yankee
M	Mike	Z	Zulu

Keep your call short. Speak clearly at a moderate pace. Talk loud enough so that the transmitter's power or RF output meter kicks up to the proper points. If you need to increase the microphone gain, use the transmitter control; don't shout. If you need less gain, turn the transmitter control down instead of whispering. Keep the microphone grille facing your mouth at a slight angle, two or three inches from your face.

Once you've finished transmitting, wait a few seconds, and if the other station doesn't respond, transmit your offer again, exactly the same as you did the first time. After that, you'll know whether or not the other station's operator can hear you (or wants to talk with you).

Sending CQ

Before you send CQ on any frequency, you should listen for at least three or four minutes to make reasonably sure that no one is having a QSO there. If it looks clear, ask if the frequency is in use, and identify your station, like this:

Is the frequency in use? This is W1GV.

Wait another half minute or so, and if nobody tells you that the frequency is in use, you can go ahead and send your CQ. Here's how I do it:

CQ, CQ, CQ, this is W1GV.
CQ, CQ, CQ, this is W1GV.
CQ, CQ, CQ, this is W1GV. Whiskey one golf victor.
Standing by!

When I say these things at my leisurely western pace, it takes me 25 seconds. Wait a half minute or so, and then repeat your CQ if nobody answers. Then wait again, call again, wait again, call again, as many times as you like, or until someone answers or tells you to go away!

An Old Timer's Quirk

If you want to make a DX contact, listen around until you hear a DX station sending CQ or ending a QSO, and then call that station. I don't recommend that you send CQ DX. I never do that. For one thing, my station is nothing close to an RF powerhouse. But even if I had a 1500-watt linear and a four-element Yagi up 100 feet, I personally would never send CQ DX. Something about it strikes me as vulgar. But maybe that's one of my many old-school hang-ups. In my view, a stateside operator should play a hunter's role, and DX should be their game!

Carrying on a QSO

The length of time that you spend in a QSO depends on the operating environment, as well as your mood. If you're working a DX station that lots of other hams are waiting to contact, or if you're in a contest, you won't get personal and the whole contact will take a few seconds. If you're "chewing the rag," however, you might go on for hours! But even when you are involved in a long, conversational QSO, don't get into extreme monologues, any more than you would in a face-to-face conversation (also known as an "eyeball QSO").

In any case, you must identify your station at the end of the contact, and at least every 10 minutes during the course of the contact. If your QSO is taking place for the benefit of a third party and it also happens to be an international communication, then you must also identify the other station. Various other rules exist for special situations. The ARRL operating manual covers some of these, but if you want the whole story straight from the lawmakers' pens, you should get a copy of the FCC regulations for radio hams.

> **Tip**
>
> Table 8-3 lists a few of the expressions you can expect to encounter in your voice-mode ham radio operation.

TABLE 8-3   Some Voice-Mode Expressions in Ham Radio Jargon

Expression	Meaning
afterburner	linear amplifier
amp	ampere; linear amplifier
barefoot	operation without a linear amplifier
beam	Yagi antenna
bird	satellite
birdie	spurious response in a receiver
boat anchor	old, antiquated, bulky equipment
brass pounder	ham who handles a lot of traffic; straight key
breaker	person trying to enter a QSO in progress
bug	semiautomatic key
call	call sign
clone	a piece of equipment that is identical to another
copy	receive; readability; written messages

TABLE 8-3 Some Voice-Mode Expressions in Ham Radio Jargon (*Continued*)

Expression	Meaning
desensing	receiver overloading by strong signal
download	receive packets from another station or from a bulletin board
DXpedition	trip to foreign land to "be DX"
eyeball	meeting in person
feedback	comments; corrections
flat	SWR of 1:1
foxhunt	hidden-transmitter search
full quieting	strong, clear FM reception
gateway	means for stations to communicate when on different frequencies and/or modes
handle	name or nickname
handy-talky; HT	handheld transceiver
header	start of a packet frame
hound	enthusiast (e.g., DX hound, CW hound)
intermod	intermodulation distortion
junk box	collection of spare parts
kerchunk	to actuate a repeater without saying anything
key; key up	to actuate a transmitter
landline	telephone
league, the	American Radio Relay League
lid	inept or discourteous operator
log off	sign off from a packet bulletin board
log on	sign on to a packet bulletin board
machine	repeater; digipeater; satellite
mike	microphone
monkey chatter	sound of SSB signal improperly tuned in

TABLE 8-3   Some Voice-Mode Expressions in Ham Radio Jargon (*Continued*)

Expression	Meaning
over	reference to dB over s-9
overhead	part of a packet that does not contain message information
paddle	key for electronic keyer
patch	connect(ion); interconnect(ion)
phone	voice communications
pileup	many stations calling a single station
rag chewing	long conversations on the radio
rig	station hardware
ritty	RTTY (radioteletype)
roundtable	contact involving three or more stations
rubber duck	shortened, flexible VHF antenna
scanner	receiver that scans for occupied or vacant channels
shack	station
skyhook	antenna
solid copy	easy-to-receive signal
splatter	spurious sidebands
swisher	person who transmits while tuning VFO
s-1 through s-9	expression of signal strength
test	contest
top band	160 meters
traffic	messages sent and received, especially on behalf of third parties
trailer	end of a packet frame, containing no information
tuner	antenna tuner; operator tuning up on frequency
upload	send packet data to another station or to a bulletin board
vox	voice-actuated operation
wallpaper	collection of QSL cards
wormhole	packet link via satellite

Ending a QSO

In SSB, ending a contact works like ending a face-to-face or telephone conversation, with the exception of the station identification requirements. You might want to say "73," which means "Best regards!," and some other form of farewell peculiar to hams, such as "See you later on down the log." If the other operator is of the opposite gender, you might also throw in "88," which means "Love and kisses" in a playful sort of way. Listen around on the SSB portions of the HF bands for a while, and you'll gradually grow familiar with the jargon and quirks peculiar to this hobby!

Operating in FM

Most ham FM operation takes place on the VHF and UHF bands, especially 2 meters and 70 centimeters. Of that activity, nearly all goes through repeaters. While FM is a voice mode like SSB, you'll behave differently on an FM repeater than you would in an HF SSB environment.

Accessing a Repeater

In the 1970s when I first got active on FM repeaters with a mobile 2-meter rig, I could almost always find a repeater and access it when I could hear it. The radio would scan the band between limits that I could preset, and then search for an occupied channel. On one occasion, I accessed a repeater and helped a family whose car had broken down at night in the country. Cell phones didn't exist then, so my communication through a repeater saved that family a lot of tedium and grief.

Today, if you can hear a repeater, you probably won't be able to use it unless you know the subaudible tone frequency for access. (Of course, we have cell phones now too, so that family probably would not need a ham like me in a similar situation now.) The tone is a steady sine-wave audio note that modulates the FM carrier at a frequency below the standard voice passband. You can obtain the tone frequency for a repeater by joining the club that operates it, or by looking it up in a database, such as the latest edition of *The ARRL Repeater Directory*.

You'll also have to know the split for the repeater that you want to use. That's the difference, either negative or positive, between the repeater's input and output frequencies. For example, a repeater might use 146.34 MHz as its input frequency and 146.94 MHz as its output frequency. That means it "hears" on 146.34 MHz and "speaks" on 146.94 MHz. In that case, you'll set your radio to transmit on 146.34 MHz and receive on 146.94 MHz. Hams call that frequency combination "34/94." Because you transmit on a frequency below the one on which you receive, hams call that split "negative" (specifically, −600 kHz). If a repeater transmits on 147.00 MHz and receives on 147.60 MHz, then you'll want to set your radio to transmit on 147.60 MHz and receive on 147.00 MHz. Users of that repeater would call its frequency combination "60/00." Because you must transmit on a frequency above the one on which you receive, you call that split "positive" (specifically, +600 kHz).

Initiating a Contact

On any repeater, you won't hear anybody send CQ unless they've never used a repeater before (or they're a non-ham who has gotten hold of a ham radio set). If you find a repeater and nobody is using it, and if you have your radio tuned to access it, you can send your call and then say "listening." That's the equivalent of CQ on a repeater. So, for example, as I roll down a Wyoming road on some crisp February afternoon, I might find a local repeater and, if no one is using it and I know the subaudible access tone frequency, say simply "W1GV listening."

Calling a Station

If you hear someone say that they're listening on a repeater, feel free to call them and have a contact! Say that station's call once, then yours once without phonetics, and then again with phonetics, just as you would do on SSB to answer a CQ.

In a repeater environment, "breaking in" is usually okay unless the ongoing communication has a priority or is for an emergency. When one of the stations ends a transmission, say your call once, clearly and without phonetics. If the other operators can hear you and don't mind your breaking in, they'll either call you straightaway, or else ask "Who's the breaker?" or something like that.

Tip

Never behave like a Citizens Band operator (CBer) on a ham radio repeater! For example, if you want to break into a QSO, don't say "Break Break Break." Simply say your call sign. Leave CB behavior to the CBers, and maintain your ham radio personality with other hams!

Carrying on a QSO

On FM, contacts proceed like they do in ordinary conversation, with the exception of the identification requirements and some technical jargon peculiar to radio hams. Keep your transmissions short and to the point. Remember that repeaters exist not only for convenience, but also (and more importantly) to serve in emergencies. You never know when someone will encounter a repeater and urgently need to use it. Leave lots of openings for such folks!

On a repeater, you should wait a few seconds after the other station finishes a transmission, and then start yours. Leave a few seconds for "breakers" who might want or need to join you.

Many repeaters have a *timeout* function. A repeater will keep putting out the carrier for a second or two after someone finishes a transmission. Then it will *drop out* (stop transmitting altogether). You'll see this event as a dive in your receiver's S-meter reading. It's a good idea to let a repeater drop out after the other station ends a transmission, and then start yours. If two hams involved in a repeater QSO don't let the machine drop out periodically, it might time out after a while anyway, bringing the QSO to an unplanned end.

Operating in CW

Since the elimination of the Morse code requirement for obtaining a ham license, CW operation has declined. Even so, a significant number of hams, myself among them, still use that mode for enjoyment. I can assure you that CW, for someone who really likes it, plays the role of an art, a science, and an addiction!

Abbreviations for CW

All experienced CW operators use abbreviations to reduce the number of characters that they have to send. These abbreviations resemble the jargon used in cell-phone texting, and in some cases, the two are identical. If you operate CW for a while, you'll grow familiar with them. Table 8-4 shows some of the most often-used ones.

> **Tip**
> During a CW contact, don't send punctuation. Separate sentences or thoughts with a break sign (DAH-di-di-di-DAH) rather than a period (di-DAH-di-DAH-di-DAH). Avoid expressions that need commas, semicolons, or other special symbols.

Calling a Station

Imagine that you hear a station calling CQ or ending a contact, and you want to start a QSO. Send the other station's call once, followed by yours once, twice, or three times, depending on how well you think the other operator will hear you, and depending on the length and "CW friendliness" of your call sign. Send at the same speed as the other station, or maybe a little slower. If the station is sending way too fast for you, consider skipping that station and trying another one, unless you want to contact it badly (a rare DX or a special events station, for example).

An Old Timer Rejoices

When I got the vanity call sign of my choice, W1GV, in 1977, I picked a character sequence that I knew other hams would easily copy on CW. If you say my call sign out loud as dits and dahs, and especially if you're a CW lover, you'll know what I mean! That sound sticks in the mind long enough for a contact to get started. It works great for cracking pileups if I send it once at 15 to 18 WPM.

TABLE 8-4   Abbreviations used in CW operation. (For Q signals, see App. B.)

Abbreviation	Meaning
aa	all after
ab	all before
abt	about
adr	address
agn	again
ant	antenna
ar	end of transmission (letters run together)
bk	break; back
bn	been; between
bt	dash; pause (letters run together)
b4	before
c	yes
cfm	confirm
ck	check
cl	going off the air
clg	calling
cq	calling any station for a contact
cud	could
dr	dear
dx	long-distance communication; foreign station
es	and
fb	fine business
fist	quality of sending
ga	go ahead
gb	goodbye
ge	good evening
gg	going
gm	good morning
gn	good night
gud	good

TABLE 8-4   Abbreviations used in CW operation. (For Q signals, see App. B.) (*Continued*)

Abbreviation	Meaning
hi	laughter
hr	here; hear; hour
hv	have
hw	how
ie	Is the frequency in use?
k	go ahead and transmit
kb	keyboard
lid	inept or discourteous operator
msg	message
n	no; numeral nine
ncs	net control station
nil	nothing
nr	number
nw	now
ob	old boy
oc	old chap
om	old man
op	operator
ot	old timer
pls	please (less common)
pse	please (more common)
pwr	power
px	publicity; press
qrq	high-speed code
r	roger (message received); decimal point
rcvr	receiver

TABLE 8-4   Abbreviations used in CW operation. (For Q signals, see App. B.) (*Continued*)

Abbreviation	Meaning
rig	station hardware
rpt	repeat; report
rst	readability/strength/tone signal report
rx	receiver; receive
sed	said
sig	signal
sk	end of contact (letters run together)
sked	regular schedule for contacts
sri	sorry
svc	service
t	numeral zero
tfc	traffic
tks	thanks (less common)
tmw	tomorrow
tnx	thanks (more common)
tt	that
tu	thank you; terminal unit
tx	transmitter; transmit
txt	text
ur	your; you're
urs	yours
vy	very
wa	word after
wb	word before
wd	word
wkd	worked
wl	well; will
wpm	words per minute
wud	would
wx	weather

TABLE 8-4 Abbreviations used in CW operation. (For Q signals, see App. B.) (*Continued*)

Abbreviation	Meaning
xcvr	transceiver
xmtr	transmitter
xtal	crystal
xyl	wife
yl	young lady ham
73	best regards
88	love and kisses

Sending CQ

If you want to start a contact but don't hear anybody else sending CQ or ending a contact, you can send your own CQ after asking if the frequency is in use. I make that query by sending "QRL?" followed by a pause and then "DE W1GV." If the frequency is occupied, you'll probably hear somebody send the single letter C (meaning "Yes"). If no one tells you that the frequency is in use, send your CQ at the speed you'd like to go in a QSO. Here's how I send a CQ under most conditions:

CQ CQ CQ DE W1GV

CQ CQ CQ DE W1GV

CQ CQ CQ DE W1GV W1GV W1GV

K

Avoid sending long CQs. If you do that, QSO candidates will get bored and tune off your frequency after a while. Would you want to hear me send the following extreme CQ?

CQ CQ CQ CQ CQ CQ CQ CQ CQ CQ

CQ CQ CQ CQ CQ CQ CQ CQ CQ CQ

CQ CQ CQ CQ CQ CQ CQ CQ CQ CQ

CQ CQ CQ CQ CQ CQ CQ CQ CQ CQ

DE W1GV W1GV W1GV

CQ CQ CQ CQ CQ CQ CQ CQ CQ CQ

CQ CQ CQ CQ CQ CQ CQ CQ CQ CQ

CQ CQ CQ CQ CQ CQ CQ CQ CQ CQ
CQ CQ CQ CQ CQ CQ CQ CQ CQ CQ
DE W1GV W1GV W1GV
CQ CQ CQ CQ CQ CQ CQ CQ CQ CQ
CQ CQ CQ CQ CQ CQ CQ CQ CQ CQ
CQ CQ CQ CQ CQ CQ CQ CQ CQ CQ
CQ CQ CQ CQ CQ CQ CQ CQ CQ CQ
DE W1GV W1GV W1GV
K K K

You'd never listen to such a broadcast all the way through from a plain old stateside station like mine, would you?

After you've finished a CQ, wait for a minute or two and then try again. Slowly tune your receiver a kilohertz or two up and down while leaving your transmitter frequency alone. Most radios have *receiver incremental tuning* (RIT) intended for this sort of situation.

Tip

Rather than sending "QRL?" to ask if a frequency is in use, experienced operators sometimes send the letters IE (di-dit, dit). However, most new hams don't know that trick, so it's grown rare. If you do hear it, now you know what it means!

Carrying on a QSO

The length and content of a CW QSO depends on the circumstances. In DXing or in a contest, it'll comprise only a signal report along with other essential information, such as your location and perhaps some contest exchange data. If you want to have a long CW conversation and the other station does too, you can keep at it all day, or as long as the band will stay open for you. Remember the identification requirements (at least once every 10 minutes), keep your transmissions down to reasonable length (a couple of minutes each time), and have fun!

Ending a QSO

When you've had enough of a CW QSO or external circumstances force you to stop, wish the other station "73" (Best regards) or maybe "88" (Love and kisses) if the other operator is of the opposite gender and you know that she or he won't be offended by such informality. Then you can send the other station's call followed by your own, and finally the letters S and K run together (di-di-di-DAH-di-DAH).

Did You Know?

Lots of CW operators conclude QSOs by sending the letter E twice (dit dit). But the formal sign-off symbol is the SK combo, di-di-di-DAH-di-DAH.

Operating in Non-CW Text Modes

You'll have an easy time operating in the non-CW text modes (RTTY, PSK31, MFSK, AMTOR, and the like) if you have CW experience. If you're not a CW operator or you are new to ham radio, read the previous section and then monitor a few QSOs on one of the non-CW digital modes to get the gist of how contacts proceed. Get the computer, the interface, and the software, and watch people send CQs, answer them, and work multiple contacts, and whatever else they do.

Tutorial Mode

I recommend PSK31 for learning how to use non-CW text in your communications. The signals are easy to find on the bands, you can almost always find a few if a band is open, and many of the people who use that mode are beginners, poor typists, or Elmers looking to help newcomers along. In addition, the tuning-aid display for PSK31, known as a waterfall, is easy and intuitive to use—and a whole lot of fun to watch and "tweak"!

Ears versus Eyes

Despite the similarities in operating procedure, a few technical differences prevail between non-CW digital modes and the good old Morse code, and these differences will affect the way you use your radio and associated peripheral equipment. For one thing, with the non-CW modes, you read the other person's stuff and send yours on a keyboard, rather than listening to audio tones and manipulating a keyer paddle or straight key. (Some CW operators use decoding software and keyboard keyers, but die-hards like me still do it the old way.) You can carry on a non-CW digital text QSO in a noisy room without headphones; on CW you'd need headphones in that same noisy room. And of course, in the non-CW digital modes, you don't have to know all the code symbols; in old-school CW, you need practice to gain proficiency.

Relative Frequency

The second distinction between CW and non-CW digital modes lies in the choices of relative frequency during communications. On CW, you can call another station on a slightly different frequency and the other party will likely hear you. In fact, rare DX station operators often prefer that you call a little above or below their frequency, not right on top of them. However, in non-CW text modes, both stations nearly always carry on their QSO on precisely the same frequency. In other words, the two signals are usually *zero beat* with each other.

Break-In (or Not)

Another difference between old-fashioned CW and the digital text modes is the fact that, with the digital modes, you can't operate full break-in as you can do with CW. The Morse code is a mark-only code; spaces are actual carrier gaps, during which your QSO partner can break in on you if your radio has the capability to hear during those gaps. In the non-CW digital modes, every transmission needs a continuous carrier; no gaps exist. The only way to obtain full break-in operation in PSK31, for example, would involve *split-band duplex* (in which one station sends on, say, 14.071 MHz and listens on 21.071 MHz, and the other station sends on 21.071 MHz and listens on 14.071 MHz). Most non-CW digital text mode operators don't want to go to the trouble to set up their rigs to work that way.

Tuning in a Signal

Yet another distinction between CW and non-CW digital text involves the tuning procedure. With old-fashioned Morse code, you can hear a signal and decode it in your head, no matter where it happens to show up in your passband. You might prefer a certain range of audio frequencies, such as 600 Hz to 700 Hz; but if your radio can hear everything that produces beat signals between 500 Hz and 1000 Hz, then you can copy that stuff too. With PSK31, RTTY, MFSK, and similar modes, you must tune your receiver precisely so that the mark and space tones fall within the narrow audio passbands of the decoding software.

Tip

If you can type fast, and if you want your PSK31 signal to come out at the highest possible speed, send everything in lowercase. Ignore the shift key except when you absolutely must use it to send a particular character. The display might look a little strange to you at first, especially when you type call signs, but sending everything in lowercase will optimize your transmitting speed (and make you look to others like an experienced PSK31 operator).

Contesting

On most weekends, one or more of the HF ham bands hosts a contest. Some contests are organized by region or locality; others involve specialties, such as CW or RTTY or 160 meters only. Contest exchanges vary in complexity; some require only a signal report and the ARRL section in which you live, while others require contact numbers or station data along with the signal report and your location or section. All contests encourage you to make as many contacts as you can, in as many different places as you can, within a set period of time, usually 24 to 36 hours.

An Old Timer Remembers

As a teenage ham in the 1960s and early 1970s, my favorite contest was the *ARRL Field Day*, held on the last full weekend in June. That contest involves portable operation using emergency stations and power sources. I loved "pulling all nighters" on CW from Assissi Heights in Rochester, Minnesota, swatting June bugs in some years and working between thundershowers in other years, sometimes freezing and sometimes sweating. No matter how bad conditions got, I always had a great time.

You Be the Prey

When you have a well-engineered antenna system and a full-legal-limit amplifier, you'll be able to "run a frequency" in a contest, sometimes for quite a while. It's fun to act as "game in a hunt," especially if you make for "tasty game"! (I live in a sparsely populated ARRL section, a characteristic that makes me an especially desirable target in contests.) When you have a "big rig," however, don't act like a bully with it. Listen for a couple of minutes and then, if you hear nothing, ask if the frequency is in use before you start sending CQs.

In CW, you can send "CQ TEST," and people will know that you're looking for QSOs in whatever contest prevails at the time. Send at a reasonable speed, such as 18 to 20 WPM. That way, you won't scare away potential contacts. While you might impress people by zipping along at 35 or 40 WPM, such show-offishness will spook less experienced operators, even if your signal is strong enough that it would make for easy copy otherwise. In SSB and other voice modes, say "CQ CONTEST" three times, then your call sign once without phonetics, then once with phonetics, and then say "standing by."

Once you get hold of a certain frequency, you can stay on it until it "grows stale" and you don't get many calls anymore, or until the band goes out, or until you collapse from exhaustion, or until the contest ends, whichever comes first. But keep in mind the fact that, in most contests, you'll get more points on the average if you make them on as many different bands as possible, and not all on one or two bands. (A few single-band events do exist, such as the ARRL 160-meter and 10-meter contests, held every year in December.)

> **Tip**
>
> Before diving into any contest, listen to some exchanges between other stations so that you get a good sense of the way things should go. Then hop in, and for a while, let your life become a scenario of "monkey see, monkey do, monkey be fast"!

You Be the Hunter

If your station has modest antennas and/or lacks a full-legal-limit amplifier, you'll find it difficult to "hold a frequency" in a contest. You can always try to capture a frequency, sending "CQ TEST" or "CQ XXX" (where XXX stands for the contest in question, such as "SS" for "Sweepstakes" or "FD" for "Field Day"), but unless you have a powerhouse station, you should expect to make more contacts if you seek out and search for stations sending CQ, and then respond to them.

In CW, transmit at a reasonable speed so that you need send your call and exchange data only one time. If you send everything at 18 WPM once and the other station gets it, you'll get at least as many contacts over time as you would get if you sent everything at 36 WPM twice. You might even go faster in effect at 18 WPM because it takes extra time whenever the other station has to ask you to repeat something. I'm lucky in CW contests and other CW competitive operating venues because my call sign is easy to copy on CW, and I hardly ever have to repeat it. For signal reports, always send "5NN" which is CW shorthand for "599," even if the other station has a rotten signal.

In SSB or other voice modes, speak clearly at a moderate pace using phonetics when necessary. Stick with the standard phonetics as listed in Table 8-2. When ending a transmission, say "Over?" as if asking a question. To confirm receipt of the other station's data, say "QSL." If you must ask someone to repeat certain data, request in plain language that they send it again.

> **Tip**
>
> It takes more work to play the role of the hunter than it does to play the role of the prey, but hunters have more control over whom they contact. So they can, if they learn to hunt well, gather more sections than their more powerful counterparts who must snare their contacts as they come, from wherever they come. That way, even though a hunter station will probably make fewer contacts than a prey station, the hunter can make up for that shortfall in part by getting more multipliers.

Working DX

One of the most intriguing aspects of ham radio is the fact that you can communicate, independently of humanmade infrastructures, with people all over the world. Hams call this practice *DXing*, where "DX" stands for "distance." But DXing involves more than maximizing the length of a communications path, which can't exceed 20,000 kilometers (half the earth's circumference) in any case. True DXing means working hams in *lots* of different recognized nations, no matter how near or distant. Many hams have confirmed contacts in more than 100 countries, thereby qualifying to join the ARRL's *DX Century Club*, and some hams have made QSOs in more than 200 countries.

Are You Astute?

You might consider moonbounce a technical exception to the 20,000-kilometer path limit because when you make a contact in that mode, the signal must travel 800,000 kilometers to the moon and back regardless of the location of the other station! But moonbounce and DXing are entirely different when it comes to operating practices and attitudes.

Choosing a Band

When you want to make DX contacts on HF, you'll have the best luck if you choose the optimum band. As a general rule, most daytime DX happens at 20 meters and shorter wavelengths, while most nighttime DX happens at 30 meters and longer wavelengths. But exceptions occur, notably on 30 meters and 20 meters. Sometimes you can work great DX on 30 meters during the day, and sometimes you can do it on 20 meters or even 15 meters at night.

Because spring and summer offers days with more sunlit time than dark time, the bands at 14 MHz and up will more likely yield good DX during those months. Conversely, because autumn and winter have days with more dark time than sunlit time, the bands at 10 MHz and below will usually work better for DX during those months. Usually. Exceptions occur, sometimes with spectacular openings on frequencies you wouldn't think could produce any contacts at all, let alone for DX.

An Old Timer Remembers

In March 2014, still technically (and weatherwise) wintertime, I tested my mobile CW radio on 10 meters and worked DX all over the world from Wyoming and Montana. Only 60 watts of output and a quarter-wave antenna, comprising a chopped-off CB whip, did the job so well that I remembered, once again, why hams call 10 meters the "magic band"!

After taking advantage of the above-mentioned guidelines as a starting point, you should listen on all the HF bands, whether it's winter or summer or day or night, and find the one with the most evidence of DX activity. You might also base your decision on other factors, such as the general noise conditions at your location (14 MHz has horrible humanmade noise at my QTH, while 10 meters has almost none). You can't work stations that you can't hear, and humanmade noise from rogue electrical appliances has proven the bane of myriad DXers' lives.

Looking for DX

The classic signature of a DX station's presence is a so-called *pileup* of stateside (United States) stations calling, usually sending their own call sign once or twice, and nothing else. For example, if you keep hearing "W1GV" over and over at intervals of about a minute, you can have confidence that I'm calling a DX station in a pileup, especially if you can also hear numerous other stations sending their own call signs at intervals of about a minute. You might hear the DX station's signal between bursts of stateside calls if it's using the same frequency as the stations calling. Listen carefully!

In a massive pileup, so many stations might call the DX that you can't separate them. On SSB it will sound like a great unruly crowd of people shouting (which, in fact, it is); on CW, it will sound like a pack of wild animals from some alien planet, howling in tones of diverse pitch. If the pileup grows extreme, some rudeness will emerge, and you'll hear people sending their call signs repeatedly and continuously. On SSB, you might hear snide comments or inane questions, such as "Screw off!" or "Who's the DX?"

Tip

Resist the temptation to answer comments and questions from lids. Keep listening for the DX, find out if the station wants specific types of calls (such as by region, by country, or on a different frequency). Wait for your chance to pounce and snare your prey with one quick strike!

Single-Frequency Mode

In most cases, DX stations will respond to calls on, or very near, their transmitting frequency, a practice called *single-frequency mode*.

If you're working SSB and you want to work DX in single-frequency mode, tune in the DX signal until the voice sounds natural, and then make certain that your radio is not set for any sort of incremental tuning, such as receiver offset (also called *clarifier*) or transmitter offset. Then, when you respond, the DX station will clearly hear you because your station's suppressed-carrier frequency will be zero beat with the DX station's suppressed-carrier frequency.

On CW, you can respond near the DX station's frequency, say 200 Hz to 500 Hz higher or lower, and consider the mode as single-frequency. Some DX stations dislike getting calls *exactly* on their frequency, so you shouldn't zero beat the DX signal. You can also listen to the pitches of the signals from operators calling the DX, and avoid sending your call exactly on any of their frequencies. That way, the DX will more likely hear you when you call along with other stations simultaneously.

Tech Note

Most radios manufactured since about 1995 have a CW offset equal to the pitch of the sidetone as you send. So if you tune in a station so that its pitch matches the sidetone pitch, you know that you're zero beat with that station, as long as you don't have incremental tuning switched on.

In PSK and other non-CW digital modes, you should always respond to a station precisely on its frequency, whether or not it's DX. That's because most operators in those modes use simple computer programs and interfaces that automatically send on the frequency to which they're tuned, and lack the capability to search for anybody who might call them on some other frequency. If you respond on another frequency, even a slightly different one, the other operator will probably miss the fact that you're calling.

Split-Frequency Mode

When a DX station operates in *split-frequency mode*, its operator wants you to call on a significantly different frequency from the one on which she or he is transmitting. Ideally, the split should be large enough so that the receiver passbands of the DX and calling stations don't overlap even slightly. That way, you'll always be able to hear the DX station when it transmits, even if dozens of other stations happen to accidentally call at the same time. Splits rarely exceed a few kilohertz, however.

When a DX station, or some other rare station such as one involved with a special event, wants to work in split-frequency mode, it will send something like "UP" or "UP 1" on CW, and give instructions in plain language on SSB. You can take "UP" to mean "Please respond a kilohertz, or two, or maybe three, above my frequency." Of course, "UP 1" means "Please respond approximately 1 kHz above my frequency."

Tip

When in split-frequency mode, DX and other rare stations almost always request that you call on a higher frequency than the one they're sending on. Once in a while you might hear "DWN" or "DWN 1," but not often.

When You Are the DX

For the adventurous spirit, nothing in ham radio can surpass the thrill of playing the role of prey in a massive DX pileup by going on a so-called *DXpedition*. Go to a different country, preferably an exotic one, and operate from there! Then you can have a great time working thousands of stations at a pace limited only by your operating skill. Once you get on the business end of a pileup, you'll find out why rare DX stations sometimes behave in strange and seemingly neurotic ways. You'll gain a good deal of operating savvy in a hurry.

If you want to go on a DXpedition, you must make arrangements for operating privileges in the country you choose. If you think that your government throws a lot of red tape in your way, wait until you deal with a foreign country, especially one that's not entirely friendly to yours! You might have to get a special license, and even obtain permission to carry your equipment into the other country. You should also make sure that you do everything according to the rules of your own country.

Tip

If you're not adventurous enough to go to a foreign country and operate from there, you can always visit ARRL Headquarters in Newington, Connecticut, and arrange to operate their official station, W1AW, for an hour or two. You'll get a pileup that way if you choose a busy band.

Rag Chewing

Sometimes the competitive aspect leaves even the most hard-bitten contesters and DXers, and they want nothing more than to carry on a casual conversation with another ham. Other hams prefer the slow life in general, and use the radio as a form of entertainment (or loneliness mitigation). In ham radio, general conversation, especially the long-winded sort, is called *rag chewing*. You can chew the rag on CW, RTTY, PSK31, MFSK, or any other mode, but most of that activity happens in the less competitive portions of the SSB subbands.

Tip

When you chew the rag, never forget that ham radio is a public medium. Anyone with a shortwave radio can tune in and hear everything you say. No law stands in their way. Why make a fool out of yourself in front of the whole world?

Use common-sense judgment when choosing topics to talk about. If you want to discuss politics with a citizen of another country, for example, you're entering perilous territory. In fact, I personally avoid political discussions in ham radio. If you want to talk about football with a Green Bay Packer fan (such as myself), you might get razzed if you root for someone else, but you won't spark World War III. If you want to talk about the weather or some other mundane topic, of course, anything goes, except spreading stuff like false hurricane warnings.

Don't make a habit of chewing the rag on repeaters. Those machines should remain available for priority and emergency communications. While rag chews on repeater are not unknown, they take place only when a repeater would likely see no use otherwise; and the rag-chewers are always ready to allow breakers to take over when they need the repeater for something important.

Warning! Amateur Radio is intended for non-profit activities only. You can't legally use ham radio for conducting business. For example, you shouldn't negotiate a real estate contract on 20-meter CW (or any other ham band). As a general rule, stay away from anything having to do with making money for yourself. Legal issues aside, this hobby is supposed to provide a distraction from workaday stuff anyway, isn't it?

Operating with QRP

If you like challenges, low-power operation (called *QRP* after the Q signal for reducing power) will provide you with plenty. I enjoy this aspect of ham radio because it gives me a chance to test new antenna designs and sharpen my operating skills. I run 10 watts output on CW and 7 watts output on PSK31, my two favorite operating modes, from my home station.

What's True QRP?

No binding definition exists for QRP, but hard-core, low-power aficionados say that you must keep your output to 5 watts or less to qualify as a full member of the QRP fraternity. Some hams run a lot less power than that! The abbreviation "QRPp" stands for an RF power output of 1 watt or less. Believe it or not, with a good antenna system and excellent propagation conditions, you can work stations on the other side of the planet with QRPp, especially on CW. The best bands for this activity are 12, 10, and 6 meters. However, some people have made QSOs over considerable distances using less than 1 watt all the way down to 1.8 MHz.

Technical Advantages of QRP

When you operate QRP, you can use battery power to run your station, and your batteries won't be five times as heavy as you are. Today's all-solid-state radios draw almost no current on receive, and not much on transmit if you scale your power down to a few watts. A deep-cycle marine battery with 35 to 50 ampere-hours of capacity can last for a full day or two with a QRP station comprising a radio such as the TenTec *Argonaut* or equivalent.

If you have a substantial radio but can scale it down to a few watts with an RF output control, QRP offers the advantage of "no-worry" continuous carrier output for the non-CW digital modes, such as PSK31, RTTY, or MFSK. With my Icom IC-746 Pro cut back to 7 watts output, I need not fear overstressing the final amplifier, even if I transmit all day long.

With QRP, you can get away with receive-grade capacitors and inductors if you want to build antenna tuners. You never have to worry about the risk of overheating or arcing in your transmission line or tuner components, no matter how high the SWR gets. If you want to force-feed an antenna whose feed line comprises small-diameter coax with a 20:1 SWR, go ahead! You won't roast anything with 5 watts, no matter how high the SWR gets.

Finally, you never have to worry about RF exposure with true QRP of 5 watts or less. Guidelines suggest that RF fields with effective radiated power (ERP) levels that low don't constitute an issue at all. In addition, you'll rarely have angry neighbors coming to your door demanding that you stop interfering with their cell phones, computers, television sets, refrigerators, toilet paper dispensers, or whatever. Unless, of course, the mere sight of your antenna spooks someone—unfortunately a not altogether uncommon occurrence.

Personal Rewards of QRP Operating

After years of QRP operation, I've evolved to enjoy the challenges, not suffer from the limitations! If you have a stacked pair of 4-element Yagis on 20 meters, one at 60 feet and the other at 120 feet (as W1AW did when I worked there in the late 1970s), along with a full-legal-limit amplifier and a top-of-the-line transceiver, you can almost always work anything you can hear. You don't need a lot of operating skill to crack a pileup with a station like that, although if you act rude, your chances go down. Brute force makes the world small.

With only 10 watts of output on CW or 7 watts on PSK31, however, I find that I have to use all my faculties to work DX on a reliable basis. But if I do keep my operating skills sharp and use every trick I know, I can sometimes beat "big guns" in a direct faceoff! When that happens, I recall one of the reasons why I got into this hobby, and why I stick with it. Wiles make up for weakness.

When your station can't dominate a frequency or a band, you must pay attention to technical issues, such as antenna design, propagation quirks, receiver passband settings, transmitter frequency offset (as opposed to receiver incremental tuning or RIT), and the fine points of operating etiquette. It's fun to let powerhouse stations contact a DX station or special event station one after another, listen to them snare their prey for a few minutes, and then, like a mouse between the feet of an elephant herd, snag a piece of that game for your log—but only because you've done your homework.

Emergency Preparedness

Ham radio justifies its existence when catastrophes wipe out humanmade infrastructures. Amateur Radio operators can provide communications when all other technologies fail. If you're interested in getting involved in emergency preparedness using Amateur Radio, and assuming that you have a license, here's what you should do to start:

- Equip your station so that it can run from emergency power for a long time. You should have a generator with plenty of fuel on hand, batteries, and spare batteries, or a stand-alone solar or wind energy system.
- Acquire a portable 2-meter FM transceiver at the very least, and if possible, a dual-band FM radio that works on 2 meters and 70 centimeters. If you can add mobile capability, so much the better; if you can add HF coverage, you've got the best.
- Contact the ARRL's *Amateur Radio Emergency Service* (ARES) and tell them that you want to get involved, and would like specific instructions. You can find them on the Web at **www.arrl.org/ares**.
- Get in touch with the officers of your local Amateur Radio club and ask about their emergency preparedness programs. If they have none, consider starting one up yourself.
- Get a copy of *The ARRL Operating Manual* and read the chapter on disaster, public service, and emergency communications. I defer to the ARRL in this category; no one knows more about the public-service aspect of ham radio than they do.
- If you're not already an ARRL member, join up!

Did You Know?

Even if an asteroid were to strike the earth, annihilating every human being on the planet except two dedicated radio amateurs, then those two people, if well-equipped for emergency communications, could talk to each other. What would they say? That subject must remain a topic for some future book!

Schematic Symbols

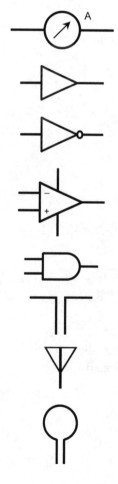

ammeter

amplifier, general

amplifier, inverting

amplifier, operational

AND gate

antenna, balanced

antenna, general

antenna, loop

antenna, loop, multiturn

battery, electrochemical

capacitor, feedthrough

capacitor, fixed

capacitor, variable

capacitor, variable,
    split-rotor

capacitor, variable,
    split-stator

cathode, electron-tube,
    cold

cathode, electron-tube,
    directly heated

cathode, electron-tube,
    indirectly heated

cavity resonator

cell, electrochemical

circuit breaker	
coaxial cable	
crystal, piezoelectric	
delay line	
diac	
diode, field-effect	
diode, general	
diode, Gunn	
diode, light-emitting	
diode, photosensitive	
diode, PIN	
diode, Schottky	
diode, tunnel	
diode, varactor	
diode, Zener	

directional coupler

directional wattmeter

exclusive-OR gate

female contact, general

Ferrite bead

filament, electron-tube

fuse

galvanometer

grid, electron-tube

ground, chassis

ground, earth

handset

headset, double

headset, single

headset, stereo	
inductor, air core	
inductor, air core, bifilar	
inductor, air core, tapped	
inductor, air core, variable	
inductor, iron core	
inductor, iron core, bifilar	
inductor, iron core, tapped	
inductor, iron core, variable	
inductor, powdered-iron core	
inductor, powdered-iron core, bifilar	
inductor, powdered-iron core, tapped	

inductor, powdered-iron
core, variable

integrated circuit, general

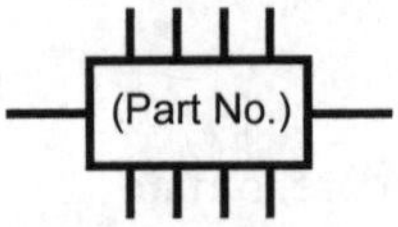

jack, coaxial or phono

jack, phone, 2-conductor

jack, phone, 3-conductor

key, telegraph

lamp, incandescent

lamp, neon

male contact, general

meter, general

microammeter

microphone

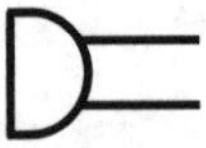

microphone, directional

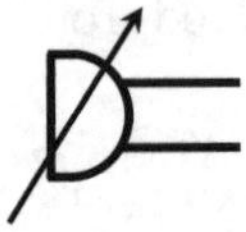

milliammeter	
NAND gate	
negative voltage connection	
NOR gate	
NOT gate	
optoisolator	
OR gate	
outlet, 2-wire, nonpolarized	
outlet, 2-wire, polarized	
outlet, 3-wire	
outlet, 234-volt	
plate, electron-tube	
plug, 2-wire, nonpolarized	
plug, 2-wire, polarized	

plug, 3-wire

plug, 234-volt

plug, coaxial or phono

plug, phone, 2-conductor

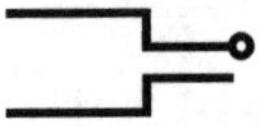

plug, phone, 3-conductor

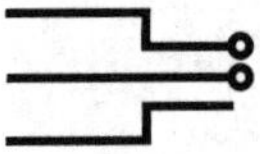

positive voltage connection

potentiometer

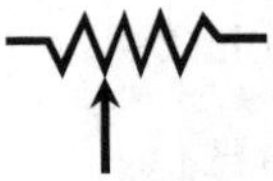

or

probe, radio-frequency

rectifier, gas-filled

rectifier, high-vacuum

rectifier, semiconductor

rectifier, silicon-controlled

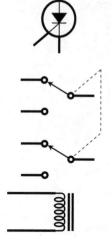

relay, double-pole,
   double-throw

relay, double-pole,
   single-throw

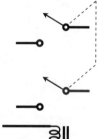

relay, single-pole,
   double-throw

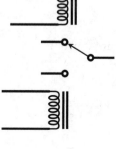

relay, single-pole,
   single-throw

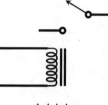

resistor, fixed

resistor, preset

resistor, tapped

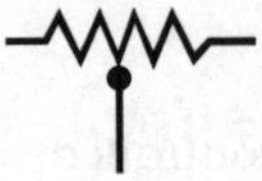

resonator

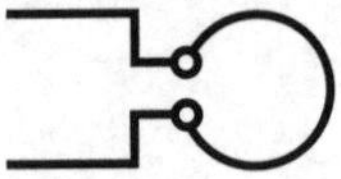

rheostat

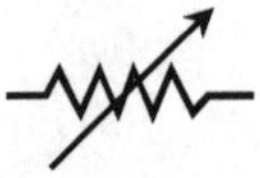

saturable reactor

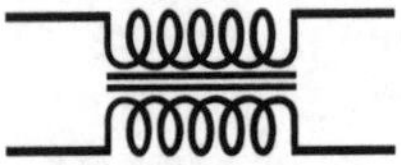

signal generator

solar battery

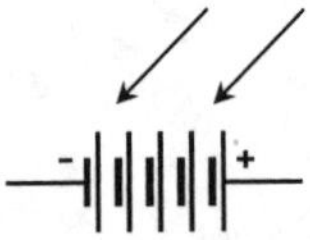

solar cell

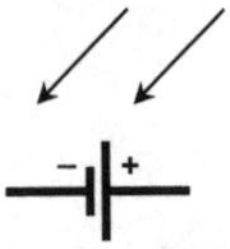

source, constant-current

source, constant-voltage

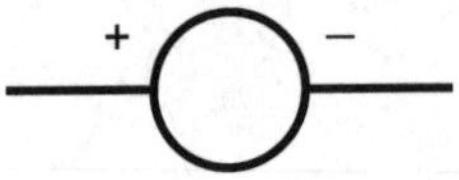

speaker

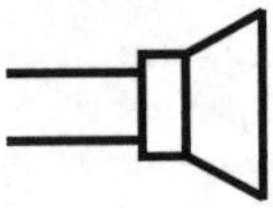

switch, double-pole,
double-throw

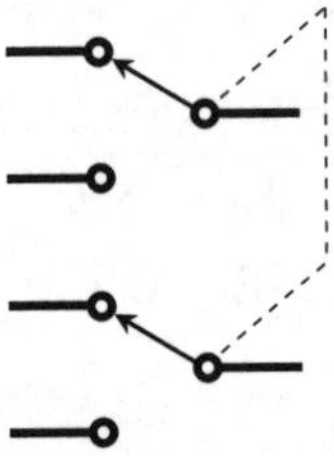

switch, double-pole,
   rotary

switch, double-pole,
   single-throw

switch, momentary-contact

switch, silicon-controlled

switch, single-pole,
   double-throw

switch, single-pole, rotary

switch, single-pole, single-
   throw

terminals, general,
   balanced

terminals, general,
   unbalanced

test point

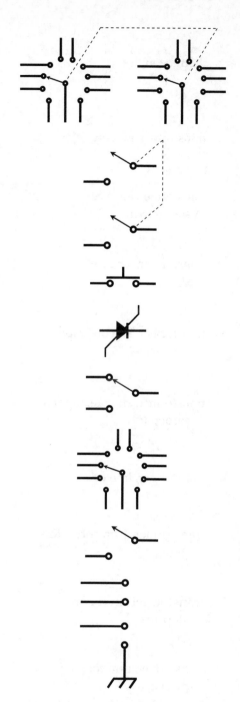

thermocouple

transformer, air core

transformer, air core,
    step-down

transformer, air core,
    step-up

transformer, air core, tapped
    primary

transformer, air core, tapped
    secondary

transformer, iron core

transformer, iron core, step-
    down

transformer, iron core,
    step-up

transformer, iron core, tapped
    primary

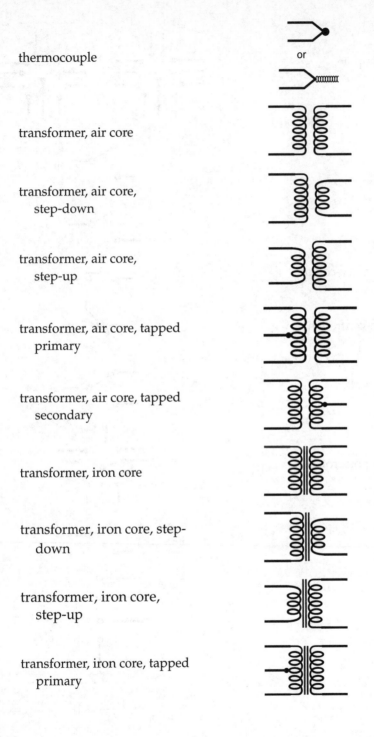

transformer, iron core, tapped
   secondary

transformer, powdered-iron
   core

transformer, powdered-iron
   core, step-down

transformer, powdered-iron
   core, step-up

transformer, powdered-iron
   core, tapped primary

transformer, powdered-iron
   core, tapped secondary

transistor, bipolar, NPN

transistor, bipolar, PNP

transistor, field-effect,
   N-channel

transistor, field-effect,
   P-channel

transistor, MOS field-effect,
    N-channel

transistor, MOS field-effect,
    P-channel

transistor, photosensitive,
    NPN

transistor, photosensitive,
    PNP

transistor, photosensitive,
    field-effect, N-channel

transistor, photosensitive,
    field-effect, P-channel

transistor, unijunction

triac

tube, diode

tube, heptode

tube, hexode

tube, pentode

tube, photosensitive

tube, tetrode

tube, triode

unspecified unit or
     component

voltmeter

wattmeter

waveguide, circular

waveguide, flexible

waveguide, rectangular

waveguide, twisted

wires, crossing. connected

or

(preferred)

(alternative)

wires, crossing, not connected

or

(preferred)

(alternative)

Q Signals for Ham Radio

These signals are used mainly by CW operators. To ask a question, send the applicable Q signal followed by a query symbol (?). To make a request, send the applicable Q signal with nothing afterward.

Q Signals for Ham Radio

Signal	Query and Statement Information
QRA	What is the name of your station? The name of my station is ___.
QRB	How far from my station are you located? I am about ___ kilometers or ___ miles from your station.
QRD	From where and to where are you going? I am going from ___ to ___.
QRG	What is my exact frequency, or that of ___? Your exact frequency, or that of ___, is ___.
QRH	Does my frequency vary? Your frequency varies.
QRI	How is the tone of my signal? The tone of your signal is: 1 (good), 2 (fair or variable), 3 (poor).
QRK	How readable are my signals? Your signals are: 1 (unreadable), 2 (somewhat readable), 3 (readable with difficulty), 4 (almost totally readable), 5 (perfectly readable).
QRL	Are you busy with ___? I am busy with ___.
QRM	Are you experiencing interference from other stations? I am experiencing interference from other stations.
QRN	Are you experiencing interference from sferics or electrical noise? I am experiencing interference from sferics or electrical noise.

Q Signals for Ham Radio (*Continued*)

Signal	Query and Statement Information
QRO	Do you want me to increase my transmitter's RF output power? Increase your transmitter's RF output power.
QRP	Do you want me to reduce my transmitter's RF output power? Reduce your transmitter's RF output power.
QRQ	Do you want me to increase my CW sending speed? Increase your CW sending speed.
QRS	Do you want me to reduce my CW sending speed? Reduce your CW sending speed.
QRT	Shall I stop transmitting? Stop transmitting. Or, I am about to go off the air.
QRU	Do you have any information for me? I do not have any information for you.
QRV	Are you ready for ___? I am ready for ___.
QRW	Do you want me to tell ___ that you are calling him/her? Tell ___ that I am calling him/her.
QRX	When will you call me again? I will call you again at ___.
QRY	What is my turn in numerical order? Your turn number is ___.
QRZ	Who is calling me? You are being called by ___.
QSA	How strong are my signals? Your signals are: 1 (almost imperceptible), 2 (weak), 3 (fairly strong), 4 (strong), 5 (very strong).
QSB	Are my signals fading in and out, or varying in strength? Your signals are fading in and out, or varying in strength.
QSD	Is my keying bad or sloppy? Your keying is bad or sloppy.
QSG	Do you want me to send more than one message? Send more than one message, or ___ messages.
QSK	Do you have full break-in capability? I have full break-in capability.
QSL	Do you confirm receipt of my message? I confirm receipt of your message.
QSM	Do you want me to repeat my message? Repeat your message.
QSN	Did you hear me on the frequency of ___ (MHz or GHz)? I heard you on the frequency of ___ (MHz or GHz).
QSO	Can you communicate with ___? I can communicate with ___.

Q Signals for Ham Radio (*Continued*)

Signal	Query and Statement Information
QSP	Will you send a message to ___? I will send a message to ___.
QSQ	Do you have a doctor or ___ on location there? I have a doctor or ___ on location here.
QSU	On what frequency (MHz or GHz) do you want me to respond? Respond on ___ (MHz or GHz).
QSV	Do you want me to send a series of V's for test purposes? Send a series of V's for test purposes.
QSW	On what frequency (MHz or GHz) will you transmit? I will transmit on ___ (MHz or GHz).
QSX	Will you listen for ___? I will listen for ___.
QSY	Do you want me to change operating frequency? Change operating frequency to ___ (MHz or GHz).
QSZ	Do you want me to send each word or group of words more than once? Send each word or group of words ___ times.
QTA	Do you want me to cancel message number ___? Cancel message number ___.
QTB	Does your word count agree with mine? My word count does not agree with yours.
QTC	How many messages do you have to send? I have ___ messages to send.
QTE	What is my bearing in azimuth degrees relative to you? Your bearing relative to me is ___ azimuth degrees.
QTH	What is your location, or that of ___? My location, or that of ___, is ___.
QTJ	What is the speed of your vehicle? The speed of my vehicle is ___ (kilometers or miles per hour).
QTL	In which direction, or toward what place, are you headed? My heading is azimuth ___ degrees, or toward ___.
QTN	When did you leave ___? I left ___ at ___.
QTO	Are you airborne? I am airborne.
QTP	Do you intend to land? I intend to land at ___.
QTR	What is the correct time? The correct time is ___ UTC.
QTX	Will you stand by for me? I will stand by for you until ___ UTC.

Q Signals for Ham Radio (*Continued*)

Signal	Query and Statement Information
QUA	Do you have information concerning ___? I have information concerning ___.
QUD	Have you received my urgency signal, or that of ___? I have received your urgency signal, or that of ___.
QUF	Have you received my distress signal, or that of ___? I have received your distress signal, or that of ___.

APPENDIX C

Ten-Code Signals for CB Radio

Ten-Code Signals for CB Radio

Signal	Query and Statement Information
10-1	Are you having trouble receiving my signals? I am having trouble receiving your signals.
10-2	Are my signals good? Your signals are good.
10-3	Shall I stop transmitting? Stop transmitting.
10-4	Have you received my message? I have received your message.
10-5	Shall I relay a message to ___? Relay a message to ___.
10-6	Are you busy? I am busy; stand by until ___.
10-7	Is your station going out of service? My station is going out of service until ___.
10-8	Is your station in service? My station is in service until ___.
10-9	Shall I repeat my message? Repeat your message.
10-10	Have you finished transmitting? I have finished transmitting.
10-11	Am I talking too fast? You are talking too fast.
10-12	Do you have visitors? I have visitors.
10-13	How are your weather and road conditions? My weather and road conditions are ___.

Ten-Code Signals for CB Radio (*Continued*)

Signal	Query and Statement Information
10-14	What is your local time, or the time at ___? My local time, or the time at ___, is ___.
10-15	Shall I pick up ___ at ___? Pick up ___ at ___.
10-16	Have you picked up ___? I have picked up ___.
10-17	Do you have urgent business? I have urgent business.
10-18	Have you any information for me? I have ___ for you.
10-19	Have you no information for me? I have no information for you.
10-20	Where are you located? I am located at ___.
10-21	Shall I call you on the telephone? Call me on the telephone.
10-22	Shall I report in person to ___? Report in person to ___.
10-23	Shall I stand by? Stand by until ___.
10-24	Have you finished your last task or assignment? I have finished my last task or assignment.
10-25	Are you in contact with ___? I am in contact with ___.
10-26	Shall I disregard the information that you just sent? Disregard the information that I just sent.
10-27	Shall I move to channel ___? Move to channel ___.
10-30	Is this action legal? This action is not legal.
10-33	Do you have an emergency message for ___? I have an emergency message for ___.
10-34	Do you have trouble? I have trouble.
10-35	Do you have confidential information? I have confidential information for ___.
10-36	Has an accident occurred at ___? An accident has occurred at ___.
10-37	Is a tow truck needed at ___? A tow truck is needed at ___.

Ten-Code Signals for CB Radio (*Continued*)

Signal	Query and Statement Information
10-38	Is an ambulance needed at ___? An ambulance is needed at___.
10-39	Is there a convoy at ___? There is a convoy at ___.
10-41	Shall we change channels? Let's change channels.
10-60	Please give me the number of your message. The number of my message is ___.
10-63	Is this net directed? This net is directed.
10-64	Are you clear? I am clear.
10-65	Do you have a net message for ___? I have a net message for ___.
10-66	Do you wish to cancel your messages numbered ___ through ___? I wish to cancel my messages numbered ___ through ___.
10-67	Shall I clear for a message? Clear for a message.
10-68	Shall I repeat my messages numbered ___ through ___? Repeat your messages numbered ___ through ___.
10-70	Have you a message for ___? I have a message for ___.
10-71	Shall I send messages by number? Send messages by number.
10-79	Shall I inform ___ regarding a fire at ___? Inform ___ regarding a fire at ___.
10-84	What is your telephone number? My telephone number is ___.
10-91	Are my signals weak? Your signals are weak.
10-92	Is my transmitted audio distorted? Your transmitted audio is distorted.
10-94	Shall I send a test transmission? Send a test transmission.
10-95	Shall I key my microphone without speaking? Key your microphone without speaking.

Ten-Code Signals
for Police and
Emergency Personnel

Ten-Code Signals for Police and Emergency Personnel

Signal	Query and Statement Information
10-1	Are you having trouble receiving my signals? I am having trouble receiving your signals.
10-2	Are my signals good? Your signals are good.
10-3	Shall I stop transmitting? Stop transmitting.
10-4	Have you received my message? I have received your message.
10-5	Shall I relay a message to ___? Relay a message to ___.
10-6	Are you busy? I am busy; stand by until ___.
10-7	Is your station going out of service? My station is going out of service until ___.
10-8	Is your station in service? My station is in service until ___.
10-9	Shall I repeat my message? Repeat your message.
10-10	Is a fight taking place at your location? A fight is taking place at my location.
10-11	Do you have a case involving a dog? I have a case involving a dog.

Ten-Code Signals for Police and Emergency Personnel (*Continued*)

Signal	Query and Statement Information
10-12	Shall I stand by until ___? Stand by until ___.
10-13	How are your weather and road conditions? My weather and road conditions are ___.
10-14	Have you had a report about a prowler? I have had a report about a prowler at ___.
10-15	Is a civil disturbance occurring at your location? A civil disturbance is occurring at my location.
10-16	Is there domestic trouble at your location? There is domestic trouble at my location
10-17	Shall I meet the complainant? Meet the complainant.
10-18	Shall I hurry to finish this assignment? Hurry to finish this assignment.
10-19	Shall I return to ___? Return to ___.
10-20	Where are you located? I am located at ___.
10-21	Shall I call ___ on the telephone? Call ___ on the telephone.
10-22	Shall I ignore the previous information? Ignore the previous information.
10-23	Has ___ arrived at ___? ___ has arrived at ___.
10-24	Have you finished your last task or assignment? I have finished my last task or assignment.
10-25	Shall I report in person to ___? Report in person to ___.
10-26	Are you detaining a subject? I am detaining a subject.
10-27	Do you have information about driver's license number ___? Here is information concerning driver's license number ___.
10-28	Do you have information about vehicle registration number ___? Here is information concerning vehicle registration number ___.
10-29	Shall I check records to see if ___ is a wanted person? Check records to see if ___ is a wanted person.
10-30	Is ___ using a radio illegally? ___ is using a radio illegally.
10-31	Is a crime in progress at your location, or at ___? A crime is in progress at my location, or at ___.

Ten-Code Signals for Police and Emergency Personnel (*Continued*)

Signal	Query and Statement Information
10-32	Is there a person with a gun at your location, or at ___? There is a person with a gun at my location, or at ___.
10-33	Do you have an emergency at your location, or at ___? I have an emergency at my location, or at ___.
10-34	Is a riot in progress at your location, or at ___? A riot is in progress at my location, or at ___.
10-35	Do you have an alert concerning a major crime? I have an alert concerning a major crime.
10-36	What is the correct time? The correct time is ___ (local or UTC).
10-37	Shall I investigate a suspicious vehicle? Investigate a suspicious vehicle at ___.
10-38	Are you stopping a suspicious vehicle? I am stopping a suspicious vehicle at ___.
10-39	Is your situation urgent? This situation is urgent; use lights and/or siren.
10-40	Shall I not use my lights and/or siren? Do not use your lights and/or siren.
10-41	Are you starting duty now? I am starting duty now.
10-42	Are you finishing duty now? I am finishing duty now.
10-43	Do you need, or are you sending, information about ___? I need, or am sending, information about ___.
10-44	Do you want to leave patrol? I want to leave patrol and go to ___.
10-45	Is there a dead animal at ___? There is a dead animal at ___.
10-46	Shall I assist a motorist at ___? Or, Are you assisting a motorist at ___? Assist a motorist at ___. Or, I am assisting a motorist at ___.
10-47	Are road repairs needed immediately at ___? Road repairs are needed immediately at ___.
10-48	Does a traffic standard at ___ need repair? A traffic standard at ___ needs repair.
10-49	Is a traffic light out at ___? A traffic light is out at ___.
10-50	Has an accident occurred at ___? An accident has occurred at ___.

Ten-Code Signals for Police and Emergency Personnel (*Continued*)

Signal	Query and Statement Information
10-52	Is an ambulance needed at ___? An ambulance is needed at ___.
10-53	Is the road blocked at ___? The road is blocked at ___.
10-54	Are animals on the road at ___? Animals are on the road at ___.
10-55	Is there a drunk driver at ___? A drunk driver is at ___.
10-56	Is there a drunk pedestrian at ___? A drunk pedestrian is at ___.
10-57	Has a hit-and-run accident occurred at ___? A hit-and-run accident has occurred at ___.
10-58	Shall I direct traffic at ___? Direct traffic at ___.
10-59	Is there a convoy at ___? Or, Does ___ need an escort at ___? There is a convoy at ___. Or, ___ needs an escort at ___.
10-60	Is there a squad at ___? There is a squad at ___.
10-61	Are personnel present in the vicinity of ___? Personnel are present in the vicinity of ___.
10-62	Shall I reply to the message of ___? Reply to the message of ___.
10-63	Shall I make a written record of ___? Make a written record of ___.
10-64	Is this message to be delivered locally? This message is to be delivered locally.
10-65	Do you have a net message assignment? I have a net message assignment.
10-66	Do you wish to cancel your messages numbered ___ through ___? I wish to cancel my messages numbered ___ through ___.
10-67	Shall I clear for a net message? Clear for a net message.
10-68	Shall I disseminate information concerning ___? Disseminate information concerning ___.
10-69	Have you received my messages numbered ___ through ___? I have received your messages numbered ___ through ___.
10-70	Is a fire occurring at ___? A fire is occurring at ___.

Ten-Code Signals for Police and Emergency Personnel (*Continued*)

Signal	Query and Statement Information
10-71	Shall I provide details concerning the fire at ___? Provide details concerning the fire at ___.
10-72	Shall I report on the progress of the fire at ___? Report on the progress of the fire at ___.
10-73	Is there a report of smoke at ___? There is a report of smoke at ___.
10-74	(No query) Negative.
10-75	Are you in contact with ___? I am in contact with ___.
10-76	Are you going to ___? I am going to ___.
10-77	When do you estimate arrival at ___? I estimate arrival at ___ by ___ (local time or UTC).
10-78	Do you need help? I need help at ___.
10-79	Shall I notify a coroner of ___? Notify a coroner of ___.
10-82	Shall I reserve a motel or hotel room at ___? Reserve a motel or hotel room at ___.
10-85	Will you, or ___, be late? I, or ___, will be late.
10-87	Shall I pick up checks for distribution? Pick up checks for distribution.
10-88	What is the telephone number of ___? The telephone number of ___ is ___.
10-90	Is there a bank alarm at ___? There is a bank alarm at ___.
10-91	Am I, or is ___, using a radio without cause? You, or ___, are using a radio without cause.
10-93	Is there a blockade at ___? There is a blockade at ___.
10-94	Is an illegal drag race taking place at ___? An illegal drag race is taking place at ___.
10-96	Is there a person acting mentally ill at ___? There is a person acting mentally ill at ___.
10-98	Has someone escaped from jail at ___? Someone has escaped from jail at ___.
10-99	Is ___ wanted or stolen? ___ is wanted or stolen. Or, There is a wanted person or stolen article at ___.

Suggested Additional Reading

American Radio Relay League, Inc., *The ARRL Handbook for Radio Communications*. Newington, CT: ARRL, revised annually. In addition to the *Handbook*, a classic in its field, I recommend all ARRL publications on topics that interest you! You can buy their books and training materials through their website at www.arrl.org.

Frenzel, Louis E., Jr., *Electronics Explained*. Burlington, MA: Newnes/Elsevier, 2010.

Geier, Michael, *How to Diagnose and Fix Everything Electronic*. New York: McGraw-Hill, 2011.

Gerrish, Howard, *Electricity and Electronics*. Tinley Park, IL: Goodheart-Wilcox Co., 2008.

Gibilisco, Stan, *Beginner's Guide to Reading Schematics*, 3rd ed. New York: McGraw-Hill, 2014.

Gibilisco, Stan, *Electronics Demystified*, 2nd ed. New York: McGraw-Hill, 2011.

Gibilisco, Stan, *Teach Yourself Electricity and Electronics*, 5th ed. New York: McGraw-Hill, 2011.

Horn, Delton, *How to Test Almost Everything Electronic*, 3rd ed. New York: McGraw-Hill, 1993.

Kybett, Harry, *All New Electronics Self-Teaching Guide*, 3rd ed. Hoboken, NJ: Wiley Publishing, 2008.

Mims, Forrest M., *Getting Started in Electronics*. Niles, IL: Master Publishing, 2003.

Shamieh, Cathleen and McComb, Gordon, *Electronics for Dummies*, 2nd ed. Hoboken, NJ: Wiley Publishing, 2009.

Silver, H. Ward, *Ham Radio for Dummies*, 2nd ed. Hoboken, NJ: Wiley Publishing, 2013.

Index

A

abbreviations for CW, 240–244
adjustable passband, 149
allwave receiver, 2
alternator whine, 230
Amateur Radio, overview of, 3–6
Amateur Radio Emergency Service
    (ARES), 256
amateur teleprinting over radio
    (AMTOR), 91–93
American Radio Relay League (ARRL), 3
American Standard Code for Information
    Interchange (ASCII), 17, 80
amplified automatic level control
    (AALC), 99
amplitude modulation (AM), 18–20
analog-to-digital (A/D) conversion,
    93–94
antenna:
    broadside, 211–212
    coaxial, 202–203
    corner reflector, 217
    dipole, 199–200
    dish, 215–216
    effective radiated power of, 207
    efficiency of, 197–199
    end-fire, 210–211
    for shortwave listening, 54

antenna (*Cont.*):
    forward gain of, 208–209
    front-to-back ratio of, 209
    front-to-side ratio of, 209–210
    half-wave, 199–201
    helical, 216–217
    horn, 215
    indoor transmitting, 225
    J pole, 201
    longwire, 211
    loop, 60–62, 203–205
    loopstick, 204
    parasitic, 212–214
    phased, 210–212
    power gain of, 207–208
    quad, 214
    random wire, 60–61
    vertical, 201–203
    Yagi, 213–214
    zepp, 200–201
antenna tuner (transmatch), 54, 144–145,
    154–156, 184–185, 221
antipode, 50
anti-VOX system, 98
ARRL Operating Manual, 227, 256
ASCII, 17, 80
aspect ratio, 103
atmospheric noise, 171–172

audio filtering, 36
audio-frequency-shift keying (AFSK), 79
audio passband filter, 149
Aurora Australis, 13
Aurora Borealis, 13
auroral propagation, 13–14
automatic frequency control (AFC), 86
automatic gain control (AGC), 153
automatic level control (ALC), 87, 98
automatic noise limiter (ANL), 149
automatic repeat request (ARQ), 88, 92
auto-tune function, 148
average forward current, 158

B

backfeed, 169–170
backward error correction (BEC), 92
backup generators, 165–170
balanced modulator, 21
ball mount, 187
balun, 221
bandpass filter, 36, 156
band scanning, 148
bandspread control in receiver, 43
baseband spectrum monitor, 67
basement as ham station location, 142–143
baud, 79
Baudot code, 79–80
beat-frequency oscillator (BFO), 27
binary phase-shift keying (BPSK), 83–84
birdies, 28
bits per second (bps), 76
blanking pulse, 104–105
bridge rectifier circuit, 159–160
broadband amplifier, 154
broadside array, 211–212
bulletin-board system (BBS), 95–96

C

capacitive coupling, 207
capacitor-input filter, 161
carbon-monoxide detector, 168–169

carrier, unmodulated, 19
ceramic filter, 150
Cassegrain dish feed, 215–216
channel scan, 148
characteristic impedance, 218, 220–221
choke-input filter, 161
circuit efficiency, 19
circular polarization, 10
Citizens Band (CB), 6–8
clarifier, 251
closed repeater, 102, 181
coaxial antenna, 202–203
coaxial cable, 218–219
code practice from W1AW, 76
coherent communications, 38
combustion generators, 165–170
contesting, 247–249
Continental code, 16
continuous receive coverage, 147
continuous waves (CW), 3, 75–79, 240–246
conventional dish feed, 215–216
converters for VLF, 63–64
corner reflector antenna, 217
coronae, 173
cosmic noise, 170
counterpoise, 206–207
current surge in antenna, 156

D

data redundancy in MFSK, 88
dead band delusion, 228
dead spot, 11
delta tuning, 149
desensitization, 150
detector:
    for AM, 32
    for CW, 32
    for FM, 33–34
    for FSK, 32
    for PM, 33–34
    for SSB, 35–36

deviation in FM and PM, 23–24
dielectric, 218–219
DigiPan software, 66–67, 84–86, 157
digipeater, 94–95
digital frequency display, 147
digital interface equipment, 156–157
digital-to-analog (D/A) conversion, 93–94
dipole antenna, 199–200
direct-conversion receiver, 27–28
directivity plots, 208
director in parasitic array, 213–214
dirty electricity, 65–72
discriminator, 34
dish antenna, 215–216
distortion, in AM, 18–19
dots and dashes, 3
double-conversion receiver, 28
double sideband (DSB), 21
dual-diversity reception, 37–38
dual VFOs, 148
duplexer, 102
DX Century Club, 250
DXing, 250–253
dynamic range of receiver, 26–27, 151

━━ E ━━
E layer, 12
earth-ionosphere waveguide, 58
earth-moon-earth communications, 15
effective ground, 208
effective radiated power (ERP), 207
efficiency of antenna, 197–199
efficiency of circuit, 19
electrical ground, 205–206
electrical system for fixed ham station, 139–142
electromagnetic fields, 8–9
electromagnetic interference (EMI), 205
electromagnetic pulse (EMP), 175–176
electromagnetic spectrum, 8–9
elliptical polarization, 10

emergency preparedness, 256
end-fire array, 210–211
envelope compression, 99
envelope detection, 32
external noise, 170–176
extremely high frequency (EHF), table, 9

━━ F ━━
F layer, 12
F1 layer, 12
F2 layer, 12
facsimile (FAX), 106–107
fast-scan television (FSTV), 103–105
Federal Communications Commission (FCC), 6
feed lines, 217–224
fidelity, 20
Field Day contest, 194
first floor as ham station location, 143–146
fixed-station ham radios, 146–152
flywheel tuning, 147
forward error correction (FEC), 88, 91
forward gain of antenna, 208–209
four-wire line, 220, 222
frame, in television image, 103
freebanders, 7
frequency display, digital, 147
frequency-division multiplexing, 38
frequency hopping, 39
frequency modulation (FM), 22–24, 100–103, 238–240
frequency multiplier, 24
frequency-shift keying (FSK), 17–18, 79
frequency sweeping, 40
frequency synthesizers, 23
frequency versus wavelength, 44
front end of receiver, 26, 31, 150
front-to-back ratio of antenna, 209
front-to-side ratio of antenna, 209–210
full break-in operation, 148
full quieting, 231

full-wave rectifier circuits, 159–160
fundamental frequency, 68

═══ G ═══

galactic noise, 170
gallium-arsenide field-effect transistor (GaAsFET), 27
generators, backup, 165–170
geodesic arc, 50
geomagnetic storm, 57
great circle route, 50
grey line, 51–52
ground:
    electrical, 205–206
    radio-frequency, 145–146, 206–207
ground loops, 206
ground-mounted vertical antenna, 202
ground-plane antenna, 202–203
ground systems, 205–207

═══ H ═══

half-wave antennas, 199–201
half-wave rectifier, 159
Hallicrafters SX-130 receiver, 43–44
ham radio frequency bands, 44–54
    graph of, 132
    proposed new, 137
    shorter than 2 meters, 135–137
    table of, 53
    2 meters, 134–135
    6 meters, 132–133
    10 meters, 130–131
    12 meters, 128–130
    15 meters, 126–128
    17 meters, 124–126
    20 meters, 123–124
    30 meters, 122
    40 meters, 119–121
    60 meters, 118–119
    75 meters, 115–117
    80 meters, 115–117
    160 meters, 113–114

ham radio license classes:
    Advanced, 112–113
    current, 109–112
    discontinued, 112–113
    Extra, 111–112
    General, 111
    Novice, 112
    Technician, 109–111
    Technician Plus, 112
ham radio license renewal, 112
hams, definition of, 2
HamScope software, 67, 84–86, 157
handshaking, 88
harmonic suppression, 152
harmonics, 66, 68–70, 87, 152
heavy portable operation, 184–185
helical antenna, 216–217
heterodyne detection, 32
HF mobile operation, advantages of, 181
HF portable operation, advantages of, 189–190
high-frequency (HF)
    definition of, 1
    table, 9
horizontal polarization, 10
horn antenna, 215

═══ I ═══

idle insertion, 89
ignition noise, 174–175, 180
image rejection, 151
images, 28, 151
impedance bridge, 155
impulse noise, 174
indoor transmitting antenna, 225
inductive loading, 179
intelligibility, 19
intermediate frequency (IF), 28, 31–32
intermodulation distortion (IMD), 30, 86–87
intermodulation spurious response attenuation, 151

International Morse Code, 16
ionosphere, 11–13
isolation switch, 169–170

J

J pole antenna, 188, 201
jamming signal, 53
junction field-effect transistor (JFET), 30

K

key, telegraph, 77
keyboard keyer, 77–78
keyer, electronic, 77–78, 149

L

ladder line, 220–221
libration fading, 15–16
license classes, ham radio:
    Advanced, 112–113
    current, 109–112
    discontinued, 112–113
    Extra, 111–112
    General, 111
    Novice, 112
    Technician, 109–111
    Technician Plus, 112
license renewal, ham radio, 112
lid, 229–230
light portable operation, 184–185
linear amplifier, 139–140, 152–154
line-of-sight waves, 10–11
local oscillator (LO), 27
long-path signal, 50
Long Range Navigation (LORAN), 113
Longwave Club of America (LWCA), 63
longwave coverage, 147
longwave radio, 2, 57–64
longwire antenna, 211
loop antenna, 60–62, 203–205
loopstick antenna, 204
loss resistance in antenna, 179, 198–199
lower sideband (LSB), 19–21, 96–97

lowest usable high frequency (LUHF), 59
low frequency (LF), table, 9
low-frequency experimental radio (lowFER), 61–63
lowpass filter, 156

M

magnetic mount, 180
main tuning control in receiver, 43
mark frequency in FSK, 32
maximum usable frequency (MUF), 13, 55–56
maximum usable low frequency (MULF), 59
mechanical filter, 150
medium frequency (MF), table, 9
memory, programmable, 148
meteor scatter propagation, 14–15
mixer, 27
mixing products, 30
MMTTY software, 81–82
mobile antenna considerations, 179
mobile band options, 177–182
mobile operation, 5–6, 177–184
mobile power options, 182–184
modem, 17, 76
moonbounce propagation, 15
Morse code, 3, 16–17, 75–79, 240–246
Morse code requirement, past, 77
multimeter, 140
MultiMode software, 83
multiple-frequency-shift keying (MFSK), 88–90
multiplexing, 38–39

N

narrowband frequency modulation (NBFM), 23–24, 101
negative feedback, 98
node, in packet radio, 94–95
noise, external, 170–176

noise blanker, 148,173
noise-canceling antenna, 70–72
noise figure, 27, 150–151
noise limiter, 173
noise quieting, 151
non-CW text modes, operation in, 246–247
notch filter, 149–150

■ O ■

open dipole antenna, 199–200
open repeater, 102
open wire, 219–221
operation of a ham station:
    AMTOR, 246–247
    contest, 247–249
    CW, 240–246
    DX, 250–253
    FM, 238–240
    MFSK, 246–247
    PSK31, 246–247
    QRP, 254–255
    rag-chew, 253–254
    RTTY, 246–247
    SSB, 232–238
optical scanner, 106
overmodulation, 98

■ P ■

packet radio, 93–96
packet radio bulletin board system (PBBS), 95–96
packets, 88, 93–96
paddle for keyer, 77–78
panoramic reception, 149
parallel-wire line, 219–222
parasitic arrays, 212–214
parasitic element, 212–213
peak inverse voltage (PIV), 158–159
penumbra of sunspot, 54–55
phase cancellation, 175
phase modulation (PM), 22–24
phase opposition, 11

phased arrays, 210–212
phase-shift keying (PSK), 67, 83–87
phasing harness, 210
phasing system, 210
phonetic alphabet, 232–233
picket fencing, 11
pirates, 7
polarization of EM wave, 10
portable antenna considerations, 186–187
portable band options, 184–190
portable operation, 184–193
portable power options, 190–193
portable whip antenna, 188
power adapter plug, 184
power gain of antenna, 207–208
power-line noise, 175, 180
power supplies, 158–165
power-supply filtering, 160–161
power transformers, 158
power transistor, 162
preamplifier, 30, 147–148
precipitation noise, 172–173
precision readout, 147
preselector, 31
product detector, 35–36
programmable memory, 148
propagation of radio waves, 10–16
pulse-amplitude modulation (PAM), 24–25
pulse-code modulation (PCM), 25–26
pulse-duration modulation (PDM), 24–25
pulse-interval modulation (PIM), 25
pulse-width modulation (PWM), 24–25
push to talk (PTT), 97–98

■ Q ■

Q multiplier, 150
QRO mobile operation, advantages of, 182–183
QRO portable operation, advantages of, 191–192

QRP mobile operation, advantages of, 183–184, 254–255

QRP portable operation, advantages of, 192–193, 254–255

quad antenna, 214

quarter-wave vertical antenna, 201–203

▬ R ▬

radials in antenna system, 202–203, 206–207

radiation patterns, 208

radiation resistance, 154, 179, 197–199

radio-frequency (RF) carrier, 18

radio-frequency (RF) ground, 145–146, 206–207

radio-frequency (RF) spectrum, 8–9

radio noise, 46

Radio Shack stores, 73

radioteletype, 17–18, 79–83

radio-wave propagation, 10–16

rag chewing, 253–254

random wire antenna, 60–61

raster, 103

ratio detector, 34

receiver incremental (or offset) tuning (RIT), 148, 251

rectangular response, 148

rectifier diodes, 158–159

rectifier circuits:

full-wave, 159–160

half-wave, 159

repeater, 101–103, 180–181, 238–239

rectangular response, 32

reflector in parasitic array, 213–214

regenerative receiver, 42

regulator chip, 162

retracing, 104

reverse-bias voltage, 159

ribbon line, 220, 222

RigBlaster Pro interface, 81, 157

ringing in audio filter, 36

ripple suppression, 161

RST system, 230–232

▬ S ▬

Saint Elmo's fire, 173

selective squelching, 37

selectivity of receiver, 26

semi break-in operation, 148

sensitivity of receiver, 26–27

sferics, 46, 171–172

shape factor, 31, 148, 152

short-path signal, 50

shortwave band, definition of, 1, 41

shortwave broadcast bands, 44–54

graph of, 53

table of, 45

11 meters, 52

13 meters, 51

15 meters, 51

16 meters, 50

19 meters, 50

22 meters, 49

25 meters, 49

31 meters, 48–49

41 meters, 48

49 meters, 47–48

60 meters, 46–47

75 meters, 46

90 meters, 46

120 meters, 45–46

shortwave listening (SWL), 41–57

shortwave radio, definition of, 1

shortwave receiver, 2

signal-plus-noise-to-noise ratio, 26

signal reporting, 230–232

signal-to-noise ratio, 26

simplex, 101

simplex teleprinting over radio (SITOR), 91

SINAD (signal to noise and distortion), 151

single-conversion receiver, 28–29

single sideband (SSB), 7, 20–22, 96–100, 232–238

skip zone, 47

sky waves, 11–13

slope detection, 33
slow-scan television (SSTV), 105–106
small backup generators, 165–170
small loop antenna, 60–62
solar flares, 56–57
solar flux, 55, 171
solar noise, 171
space frequency in FSK, 32
spectral display, 19, 85–86
speech clipping, 100
speech compression, 98–99
splatter, 86–87, 99–100
split-band duplex, 247
split-frequency mode, 252
squelch sensitivity, 151
squelch threshold, 36–37
squelching, 36–37
solar flares, 14
solar noise, 15
sporadic-E propagation, 12
spread spectrum, 39–40
standing-wave ratio (SWR), 153, 185,
    223–224
standing waves, 223–224
step-down transformer, 158
step-up transformer, 158
sunspot cycle, 55–56
sunspots, 54–57
superheterodyne receiver, 28–29
super high frequency (SHF), table, 9
surface waves, 11
surge protector, 140
SWR meter, 185, 189
synchronized communications, 38

T

telegraph key, 77
teleprinter, 79–80
television:
    fast-scan, 103–105
    slow-scan, 105–106
terminal node controller (TNC), 93

time-division multiplexing, 38
time-shifting communications, 93
tone squelching, 102
transceiver for ham radio, 3–5
transformer, power, 158
transient suppressor, 140, 163–164
transmatch, 54, 144–145, 154–156,
    184–185, 221
transmission lines, 217–224
transmit/receive (T/R) switch, 88
transmitter offset, 251
tropospheric bending, 13
tropospheric ducting, 13
tropospheric propagation, 13–14
tropospheric scatter (troposcatter), 13
tubular line, 220–221
twinlead, 222

U

UHF mobile operation, advantages of,
    180
UHF portable operation, advantages of,
    187–189
ultra high frequency (UHF), definition
    of, 2
umbra of sunspot, 54–55
uninterruptible power supply (UPS),
    163
unmodulated carrier, 19
upper floors as ham station location, 146
upper sideband (USB), 19–21, 96–97

V

velocity factor, 199
vertical antenna, 201–203
vertical polarization, 10
very high frequency (VHF):
    definition of, 2
    table, 9
very low frequency (VLF), table, 9
voice-mode expressions in ham radio,
    235–237

voltage doubler, 160
VHF mobile operation, advantages of, 180
VHF portable operation, advantages of, 187–189
VibroKeyer Deluxe, 78
VLF converters, 63–64
voice-operated transmission (VOX), 97–98
voltage regulation, 162
volt-ohm-milliammeter (VOM), 140

W

waterfall display, 67, 85–86
waveguide, 223

waveguide propagation, 58
wavelength versus frequency, 44
weight control for CW, 150
wideband frequency modulation (WBFM), 24, 101
words per minute (wpm), 16, 75–76
working DX, 250–253
W1AW code practice, 76

XYZ

Yagi antenna, 213–214
Zener diode, 162
zepp antenna, 200–201
zero beat, 27, 247